AF324738

Nothing is Permanent, says Bhagwad Gita
implying
Only Change is Constant

The Indian military must change with shifting
geostrategic dynamics and in sync with twenty-first century
Network-centric Warfare

*We are responsible for what we are, and whatever we wish ourselves to
be, we have the power to make ourselves. If what we are now has been
the result of our own past actions, it certainly follows that whatever we
wish to be in future can be produced by our present actions; so we have
to know how to act.*

—Swami Vivekananda

Indian Military
and
Network-Centric Warfare

Indian Military
and
Network-Centric Warfare

Lt Gen (Retd) PC Katoch
PVSM, UYSM, AVSM, SC

First published 2014

ISBN 978-81-8328-362-5

Published by
Wisdom Tree
4779/23, Ansari Road
Darya Ganj, New Delhi-110 002
Ph.: 23247966/67/68
wisdomtreebooks@gmail.com

Printed in India

This book is dedicated to the Indian military, all those who are striving to take the military into the next level of netcentricity and all the future net-centric warriors who will lead India from one victory to another in future conflicts.

Content

Foreword

Information superiority, like air superiority, has now become a necessary condition for the conduct of successful military operations. Modern militaries are endeavouring to achieve it by Network-centric Warfare or Net-centric Warfare (NCW).

NCW aims at the near-instantaneous distribution of a large amount of data, which enables different units and even individuals to share a common operational picture. Networking provides a new type of advantage which results in significantly improved capabilities for sharing and accessing information. The characteristics of NCW are speed, precision, knowledge and innovation.

The relationship between information and combat is well known. However, the challenge has always been as to how it can be maximised. All military operations are conducted in three domains. Two of these—the physical domain and the domain of the mind—are well known and understood. The physical domain is where attack, defence and manoeuvre occur—on ground, sea, air or space. Elements of this domain are easy to measure, eg lethality and survivability.

The domain of the mind is where battles are won and lost. This is the domain of the intangibles: Leadership, morale, unit cohesion, level of training and experience, public opinion and so on. Key attribute of these intangibles have remained relatively constant.

The third domain is that of information. It is this domain which is now increasing combat power in a broad range of operations. It is network-centric forces that help us to share a common operational picture, resulting in a very high level of shared situational awareness.

The current military hierarchical model relies heavily on centralised control and detailed orders. These are implemented by the subordinates as best as they can. There may be some changes on account of the tactical situation, but by and large, the endeavour is to stick to laid-down plans and orders to the maximum extent. Centralised control has been an acceptable way to exercise command. With the changing nature of war, however, the importance of actions of subunits and, at times, even of individuals, has increased exponentially. This translates into adopting the *directive style of control,* which is based on effective delegation of authority. This is a direct result of appreciating that the future battlefield milieu is a mix of uncertainty and chaos. Networks are particularly suitable for this environment, as they distribute decision-making. Decision-making, thus, becomes a mix of bottom-up as well as top-down approaches.

The intention of *directive style of control* is not to replace hierarchies, but make them more attuned to initiative by subordinates and optimally use fleeting opportunities. This ability to increase combat power at tactical-levels provides operational commanders with increased flexibility to employ their forces and to generate desired effects. NCW, thus, provides commanders with an improved capability for dictating the sequence of battle and the nature of engagements, controlling force ratios and rapidly foreclosing enemy's courses of action.

There is a strong correlation between information sharing, improved situational awareness and significantly increased combat power in a network-centric force or its components, as it has the capability to share and exchange information among geographically dispersed elements of the force, whether they are sensors, shooters, decision-makers or supporting organisations. This results in a common operational picture and a common tactical picture. Better networked forces may well be the key to the future battlefield. It also needs to be highlighted that network-centric operations impact at all levels of war—strategic, operational and tactical.

Networks are complex systems that thrive on connectivity, flat organisations and peer-to-peer links. These are increasingly becoming important for providing regularity and predictability. Increased network connectivity increases communications, as they are not subject to the filtering and limits inherent in a vertically integrated hierarchy. Unlike hierarchies, networks spread the responsibility for planning and decision-making across many subordinate levels. There are drawbacks too, but in the overall context there are important benefits.

We have been developing a range of networking capabilities in all the three Services but usually with an environmental, rather than a holistic or joint focus. NCW capability requires a multidisciplinary and holistic approach, which includes the development of matching doctrine and infrastructure, restructuring and even re-engineering of some organisations. It also requires qualitatively higher knowledge and skill thresholds. The armed forces, the entire national security establishment, Defence Research and Development Organisation (DRDO), Public Sector Undertakings (PSUs) and Information Technology (IT) companies have to work closely in unison and leverage their capabilities, so that a truly Indian system emerges. There is need to take a hard look at all these issues and give impetus to this process through a well-thought-out plan.

The armed forces currently conform to an organisational philosophy of vertically integrated command and control. This system reduces uncertainty by providing a framework that specifies how individuals, subunits, units and formations should interact with each other. On the other hand, transition to a network enabled force requires an organisational system that increases the productive capacity of the subordinates by maximising individual and variable human intellectual effort.

Increase in networking is achieved by investing in networks and by educating and training soldiers who operate the network and fight another networked force. Training joint and combined forces that have compatible networking capabilities is important for the development of new tactics, techniques and procedures.

All new technologies represent a challenge to an existing social order and imply gain for some constituencies and loss for others. There is also an intra-organisational competition for resources and status. Since the results of networked technologies are likely to have their major impact on the sociology of military organisations, the greatest challenges that the armed forces will face are likely to be cultural, in the form of introducing changes in thinking and behaviour. Existing attitudes and beliefs about how warfare is conducted today may well be the biggest impediment in achieving network-centric capabilities.

The US military was the first to adapt NCW in a big way. Others, notably, the British and Australian militaries, favour the term Network Enabled Capability. This is obviously a truncated model, which may well suit the Indian military too, at least for the present.

Opportunities afforded by new technologies need to be adopted to increase military capabilities. The ability of military leaders to adapt to a 'network-centric' force is less a problem of technology than a problem of attitudes and social order. New technologies are developed to solve problems, but they also disrupt, subvert and reshape the existing order. In conservative institutions like the armed forces, this tends to be resisted, hence the importance of changing behaviour gradually.

It may not be as simple as it first appears, but it can be achieved with time and training. The ability of the Indian military to become an adaptive, versatile and flexible force depends on the capacity of subordinate formations and units to absorb and integrate new technology.

The various aspects mentioned above, as well as many more details have been covered comprehensively by the author in this book, aptly titled, *Indian Military and Network-centric Warfare*.

General Katoch needs to be complimented for including all major aspects of NCW in his book. A subject of this nature can easily get bogged down in technicalities, which may result in the book becoming a technical treatise. However, it needs to be remembered that while NCW makes use of technologies, it is essentially a new and different way of warfighting that aims at winning battles and wars in the battlefield milieu of the future. Technology, especially the electromagnetic environment, is important indeed, but it is an enabler for the armed forces to conduct military operations efficiently in a vastly changed battlefield environment.

The book maintains a balance between the requirements of technology and the need to get on with military operations so that victory can be achieved in a reasonable time frame and casualties of own troops are kept to the minimum. This has been achieved by keeping an objective approach and combining knowledge and experience of the author. General Katoch has also incorporated knowledge gleaned from specialists in different aspects of the subject. This adds to the value of the book. His research has also led him to conclusions that make the reader understand the dynamics of netcentricity, in how future military operations are likely to be conducted.

As the second decade of the twenty-first century rolls on, we are approaching an operationally-demanding future, which is likely to radically change our approach to battles and wars. I recommend this excellent book to all those who seek an understanding of the changing character of future armed conflicts. This includes the discerning non-military readers too, as the book is easy to understand and gives an insight into how the armed forces fight wars today and how they will do so in the future.

Let me conclude by stating that the road to an efficient networked force requires a multi-disciplined intellectual approach. While developing an increasingly networked force, we need to keep in mind the ethical dilemma of how much decision-making should be devolved to subordinates when that decision-making vitally affects military personnel on the battlefield. Ultimately, wars are won by those soldiers and their leaders who are technologically prepared and equipped for the battlefields of the future.

Lieutenant General (Retd) Vijay Oberoi, PVSM, AVSM, VSM
Former Vice Chief of Army Staff, Indian Army

Acknowledgements

This book would not have come about but for Brigadier Gurmeet Kanwal (Retd), former director, Centre for Land Warfare Studies and Major General Dhruv C Katoch, SM, VSM (Retd), former deputy director, Centre for Land Warfare Studies, who subsequently took over as director from Brig Kanwal. When Gurmeet, an Artillery officer and Dhruv, an Infantry officer, asked another Infantry officer like me, from the Special Forces, to write a book on this subject, it was an offer I simply could not resist. My thanks to them both and I hope I have done justice to their faith reposed in me.

I would like to acknowledge and place on record the inspiration that I drew, as many like me would have, from Lieutenant General Vijay Oberoi, PVSM, AVSM, VSM (Retd), former Vice Chief of Army Staff—a highly decorated, distinguished and respected Infantry officer who almost a decade-and-a-half ago, while heading an Army study, conceptualised Information Systems for the Indian Army, laying the basic framework for what is still a work in progress. Post superannuation, as the first director of the Centre for Land Warfare Studies, he continued to vigorously pursue Network-centric Warfare (NCW) through discussions and seminars, to assist the military to remain focused on the issue. I consider myself fortunate that he has written the foreword for this book.

I stand indebted to numerous officers of the Army, Navy and Air Force (both serving and veterans) who have given their invaluable inputs willingly in order to improve the system and forge ideas to take the Indian military to the required levels of netcentricity. Three of them who occupied very high decision-making positions, were candid enough to admit they do not understand technology while another, who had served on the Defence Procurement Board, provided an insight into the workings of the Defence forces. Serving officers obviously chose to stay anonymous, lest the system so attuned to cosy turfs and attitudinal resistance to change closes ranks against them.

A special thanks to Saikat Datta, resident editor, Delhi, of the *Hindustan Times* who has taken pains to design the cover for this book, and with whom I had co-authored an earlier book, *India's Special Forces—History and Future of Indian Special Forces*.

I thank scores of officials from the Defence Research and Development Organisation (DRDO), Public Sector Undertakings, private industry and scholars with whom I interacted while in service and continue to do so now through periodic meetings, discussions, seminars and forums. The names are a long list and their inputs invaluable, all conveying the same message that as a nation, we have it in us and that we can do it equally, if not better than any nation, provided we can get our act together.

Finally, a big thanks to my wife Kamini, son Anuj and daughter-in-law Deepika who gave me their whole-hearted support in this venture.

PC Katoch

Preface

Rapid advancements in information technology have enabled robust networking of well-informed, geographically dispersed forces, contributing to an information advantage. But this is just one aspect of Net-centric Warfare (NCW). It is often assumed that NCW refers to nothing more than a reliance on technology, which is a misinterpretation of the doctrine. Networking, enabled by information technology, needs to be meshed with overall technological advancements and combined with organisational structures, processes and above all, people—warfighters, operators, managers and commanders. NCW encompasses the entire gamut of emerging military responses in the Information Age. An NCW-capable force is robustly networked, with improved information sharing, situational awareness, collaboration, self-synchronisation, sustainability, speed of command and mission effectiveness. Lack of networking creates avoidable gaps in information availability and duplication of tasks. At critical moments of national security, there is no scope for any breaks in downtime. The last decade has seen significant improvements in sensors, high-speed digital data transfer using world wide web, optical and mobile telephony links and the ruggedisation of hardware, coupled with greater affordability. The use of cutting edge sensor technology has ensured that results remain unaffected by bad weather and light conditions. The important thing is that precise data and imagery for quick and accurate decision-making are available on call, ensuring battlefield transparency and situational clarity even under the most trying circumstances.

The Indian military is, perhaps, a decade-and-a-half away, if not more, from emerging as a truly NCW-capable force. The lack of political direction and impetus, in terms of a Revolution in Military Affairs, has also contributed to this slow progress. Information is not acknowledged as a strategic asset in the true sense, which has prevented the adoption of a top-down approach. Therefore, tri-Service netcentricity remains a casualty and there is a host of disparate systems, which are being integrated at an excruciatingly slow pace. We need to usher in a Revolution in Military Affairs (RMA), keeping in mind the current and emerging threats that we face.

Introduction

Netcentricity in the Indian military continues to be in a stand-alone mode amongst the three Services. Intra-Service netcentricity is higher within the Navy and Air Force, because the type of weapon systems that they operate cannot function without networking. So, netcentricity is perforce streamlined in these two Services. In the Army, netcentricity has mushroomed bottom upwards.

The lack of a tri-Services Net-centric Warfare (NCW) doctrine has resulted in a deficient tri-Service NCW architecture, which is still being worked upon. Non-integration of HQ Integrated Defence Staff (IDS) with the Ministry of Defence (MoD), HQ IDS's limited authority, coupled with operational responsibility and lack of a Chief of Defence Staff (CDS), have all contributed towards the above. Though we have doctrines for command, control, communications, computers, information and intelligence (C^4I^2), as well as Information Warfare (IW), these two spheres are components of NCW and do not constitute NCW by themselves. NCW must also encompass policies, strategies, concepts, military organisations and adjustments. To transform the Indian military into an NCW-capable force, we need a joint NCW doctrine as the starting point. The concepts developed by individual Services should flow from this joint doctrine. This will facilitate the development of a coherent tri-Service networked architecture.

Presently, the networks of the three Services are not interoperable or integrated. Neither our voice and data networks, nor our radio communications, are interoperable to the desired degree. Each Service develops networks on its own and starts thinking of interoperability at a much later stage. Common standards and protocols for the three Services have not been evolved. Finalising and adoption of standards and protocols, mutually compatible database structures, development/deployment of interfaces between systems using disparate platforms and commonality of hardware are challenges which need to be overcome. No single unifying secrecy-algorithm for the three Services has been developed either. There is an absence of knowledge management. The three Services need to closely study collaborative working, not only across the Services but also within each Service. The command and control structures will have to cater for collaborative working. Although a network-enabled environment for the Indian military would be available at the operational-level in a decade, it is the change in mindsets and absorption of technology that are likely to take up most of the time. Modern information technology permits the rapid and effective sharing of information to such a degree that 'edge entities', or those that are essentially conducting military missions themselves, should be able to pull information from ubiquitous repositories, rather than having centralised agencies attempt to anticipate their information needs and push it to them. For example, a company commander can 'pull' information in a networked environment fielding a Battlefield Surveillance System (BSS) from sensors operating under the Brigade, Division, Corps or higher level. This would imply a major flattening of traditional military hierarchies, which goes against the existing age-old hierarchal vertical military chain of command, creating an unwanted fear of loss of turf. There are challenges of a painfully labourious procedure of procuring information systems through the Defence Procurement Procedure and layers of Services and civil bureaucracies, coupled with our unaccountable and unfocused defence research and development (R&D).

The old maxim, 'the pursuit of excellence is a never-ending process' merits serious consideration here. It goes without saying that, in order to improve, it is essential to identify and incisively analyse the shortcomings and voids in the existing system. The writings in this book, therefore, need to be viewed in this context and the critique is not intended to ruffle the feathers of any Service, arm, organisation or individual. Since the Navy and Air Force are compact Services with existing netcentricity at comfortable levels, the discussions will focus on the Indian Army (IA) and tri-Service integration.

History will mark the twenty-first century as the age of networking, with individual systems plugging into larger systems, thus leading to the ultimate goal of a System of Systems. We must take full advantage of this and aim for complete NCW capability for our armed forces and the nation, to ensure the attainment of our national security objectives.

Finally, it is the interplay between Revolution in Military Affairs (RMA), NCW and command, control, communications, computers, information, intelligence, surveillance and reconnaissance (C^4I^2SR). The innovative application of technology as part of RMA, together with suitably changed doctrine and concepts, brings about a fundamental change in the character and conduct of operations, meeting twenty-first century requirements. NCW and C^4I^2SR too, are based on optimisation of technology, particularly information technology. Therefore, RMA, NCW and C^4I^2SR are so intimately interlinked that they cannot be surgically segregated into separate compartments. To that end, readers will need to bear with some repetition. However, the endeavour has been to keep it to a minimum but, to an extent, so that individual chapters remain holistic.

1

The Impact of
Technology on War

[1]Incredible human effort has gone into extending the range and lethality of weapons over the years, and advancements in technology have revolutionised warfare already. By 2025, technology will have gone to the next step or perhaps, the next to next step. With the continuing volatility in India's neighbourhood, we may be faced with heightened threats in future through a spectrum of conflict, particularly in the asymmetric spheres, along with the activated battlefields of space and cyberspace. We need to examine how technology will impact future warfare; what our technological voids and weaknesses are and what initiatives we need to take in order to enable India to gain its rightful place in the comity of nations.

Current Impact

[2]Technology gave birth to the term 'modern war'. U-boats were first used in World War I (WWI), as were airplanes and tanks. A new

---

[1]Toffler, Alvin and Toffler, Heidi, *War and Anti-War: Survival at the Dawn of the 21st Century*, Warner Books, New York, 1995.

[2]Creveld, Martin van, *Technology and War: From 2000 BC to the Present*, Free Press, New York, 1991.

type of warfare emerged with poison gases like mustard gas. [3]Many developments occurred during and after World War II (WWII), with the Industrial Revolution serving as a catalyst. In WWII, the use of tanks and airplanes increased. More aerial bombings were resorted to along with the use of V2 rockets. The development of the atomic bomb ushered in the nuclear era—one could simply push a button and blast a city to the ground. The hydrogen bomb was even more potent. The element of surprise overshadowed marching into battle in phalanxes or using trenches. [4]Six decades into the nuclear age, came a growing belief that the hydrogen bomb may prove to be the sought-after war-stopper. [5]The Gulf War demonstrated a number of hi-tech weapon systems, surveillance and target acquisition and command and control systems. Presently, technology enables hi-tech wars that are short and swift. The ranges, accuracy and lethality of weapons have been enhanced considerably. Concurrently, the space and time continuum has been greatly compressed. There is an exponential increase in situational awareness and battlefield transparency as forces are shifting from platform-centric to network-centric capabilities. [6]The simultaneous handling of the strategic, operational and tactical-levels is possible. Improved battlefield transparency has, in turn, increased the need for the dispersion of forces and deception. [7]Technology has ushered in the advent of offensive cyber warfare, information dominance, space wars and

---

[3]Lawrence, Philip K, *Modernity and War: The Creed of Absolute Violence*, Martin's Press Inc, New York, 1997.

[4]Gaddis, John Lewis, *The Long Peace: Inquiries Into the History of The Cold War*, Oxford University Press, London, UK, February 1989.

[5]Boot, Max, *War Made New: Technology, Warfare, and the Course of History: 1500 to Today*, Gotham Books, New York, October 2006.

[6]Tuner, Bunyamin, 'Information Operations in Strategic, Operational and Tactical Levels of War—A Balanced Systematic Approach', http://www.dtic.mil/dtic/tr/fulltext/u2/a418305.pdf

[7]Lister, Jonathan, 'The Impact of Technology on Warfare', 2013, http://www.ehow.com/info_7844707_impact-technology-warfare.html

Effects Based Operations (EBO). But ironically, technology has also empowered the terrorists to cause more severe damage.

Future Technological Developments

Listing the likely technological advancements in military applications by, say, year 2020, will take up many pages. For the sake of better understanding, let us take a few examples. The improved assault rifles will perhaps include phased plasma guns. [8]Satellite-fuelled solar-powered tanks are being talked about. Plasma weapons, already reported in Russia, are said to focus beams of electromagnetic energy produced by laser or microwave radiation into the upper layers of the atmosphere, to defeat targets flying at supersonic or near-sonic speeds, bumping the targets out of their trajectory. Laser weapons will be mounted on land, sea and air. Our Defence Research and Development Organisation (DRDO) is developing a Laser Dazzler for police forces that will temporarily impair vision, to control unruly crowds. DRDO's Laser Science and Technology Centre (LASTEC) is developing a vehicle-mounted gas dynamic laser-based Directed Energy Weapon (DEW) system, named 'Aditya', as a technology demonstrator. A 25kW laser system is also under development to hit a missile in its terminal phase at a distance of 57 kilometre. DRDO identifies DEWs, along with space security, cybersecurity and hypersonic vehicles, as its future projects. The 'Technology Perspective & Capability Road map' of our Ministry of Defence (MoD), identifies DEWs and anti-satellite weapons (ASAT) as thrust areas over the next fifteen years. However, given DRDO's track record, how much they will actually deliver by 2025 is anybody's guess.

While there was talk of Stealth Helicopters in the recent raid by US Navy Seals in Pakistan to kill Osama bin Laden, the US has

---

[8]Sharoni, Asher H, and Bacon, Lawrenica D, 'The Future Combat System (FCS): A Satellite-fueled, Solar-powered Tank?' *Armor*, January-February 1998.

already developed a Reusable Space Plane (X-37B)—another step in weaponising space. In addition, a powerful conventional weapon (Prompt Global Strike), has been developed as an alternative to a nuclear warhead. It can travel halfway around the world from launch to target in less than thirty minutes, using a missile launch, the release of hypersonic gliders and the eventual release of 1000-pound deep penetration bombs. The revolution in communications equipment is visible in commercialised products. The narrative of technological advancements can go on endlessly. Spying, snooping and reverse engineering has given China access to the designs of the US F-I6, the B1 Bomber, the US Navy's quiet electric drive, the US W-88 miniaturised nuke (used in Trident Missiles) to name a few. China aims at parity with the US in science and technology in three or more decades. The J-20 stealth fighter has been developed in record time. Stealth helicopters and vessels should be following. An indigenous aircraft carrier under development is estimated to be twice as fast, with double the capacity to launch and land through twin decks. China's indigenous aircraft carrier may have twice the speed of existing carriers with a catamaran type of hull, greatly reducing pitching, yawing and swaying, while enabling the simultaneous launching and landing of aircraft from twin flight decks.

India's Space Vision 2025 envisages satellite-based communication and navigation systems for rural connectivity, security needs and mobile services. Imaging capability is to be enhanced for natural resource management, weather and climate change studies. The Indian Regional Navigational Satellite System (IRNSS), with its seven satellites, is on its way to deployment, and will enable the fielding of an indigenous Global Positioning System (GPS). Space missions and planetary exploration are planned to understand the solar system and the universe. The development of a heavy lift launcher and reusable launch vehicles as technology demonstrator missions, leading to Two Stage To Orbit (TSTO) and human space flight, are also planned.

Battlefield Transformation

Technology has always played a role in warfare. Improvements in military weapons throughout history have forced armies to continually adopt new fighting tactics to win battles. This is still true in the modern era, where advances in robotics and targeting systems have led to smarter weapons with deadlier payloads. [9]Just as munitions have become smarter with the military's ability to target them with increasing accuracy, they have also become deadlier. According to Fox News, the US military has some of the world's deadliest weapons in its arsenal. For example, the AC-130 aerial gunship mounts a 75 mm cannon, which is able to blow through buildings, pierce armoured vehicles and remove cover from the adversary. The high rate of fire from a craft like the AC-130 leads to a higher rate of enemy casualties. Technology in warfare has meant an increase in air support for wartime missions, as well as an increase in the numbers of unmanned aircrafts. This has meant fewer soldiers on the ground during the initial stages of a war, which translates into fewer military casualties. Fighter and bomber pilots can remove the defences of an opposing army, utilising precision munitions without needing direct ground assault. When ground troops move into the combat zone, they confront a significantly depleted fighting force. This strategy was employed by the US military during the first and second Iraq Wars to remove defensive capabilities and demoralise the existing military force.

The best combination on the modern day battlefield is of NCW and EBO. [10]EBO implies focused application of comprehensive national power that includes military power. A military that has

---

[9]Smith, Edward R, 'Effect Based Operations', http://www.dtic.mil/dtic/tr/fulltext/u2/a457292.pdf

[10]Carr, Peter, Information Technology and Modern Warfare 2013 Update, 28 February 2013, http://impactofinformationsystemsonsociety.wordpress.com/2013/02/28/information-technology-and-modern-warfare-2013-update/

acquired NCW capabilities, having shifted from attrition-style warfare to a faster and more effective warfighting style characterised by the new concepts of speed of command and self-synchronisation, will, obviously, enable optimisation of EBO. Since NCW is an enabler that transcends the physical, information and cognitive domains, conflict response becomes that much more effective throughout the complete spectrum of conflict. EBO shifts our focus from targets and damage to behaviour and the stimuli that alter behaviour. This broad multilevel interaction will form the basis of a new strategic deterrence. NCW capabilities and thinking are still evolving and India will need to take the requirements of EBO into account. Beginning with using the power of the network as best as we can, we must progress to using new information technologies to deal with the three levels of complexity at the core of operations in the cognitive domain. Developments in decision aids that handle the multiple levels of complexity will be of immense help. Such aids manage what an observer sees and demonstrate the cascading effects that a given option may produce. Decision aids can provide relative probabilities for a given behaviour in a given situation. Another area of promise is in the use of indicators to provide feedback on the effects of our actions; although the effective use of indicators will probably revolve around data mining and set up automated algorithms to spot behavioural and performance changes that constitute definitive indicators and then render the indications in terms of probabilities of a particular behaviour taking place. Considerable research is underway to incorporate human behaviour into Decision Support Systems by data mining human behaviour in various situations. The research is complex, involving complex mathematics and algorithms. It does hold promise though a time frame of its completion will remain unpredictable due to multiple uncertainties and large number of interdependent variables including individual human behaviour. Even when such a result is completed, it is unlikely that any country would like to share it, which is akin to considerable research being undertaken globally, including in India, on mind control. These Information Age capabilities could clearly play a major role in the future of EBO. The results will always be related to human behaviour, but the faster we identify the parameters of what we need, the faster they will be in the commanders' hands.

Considering the rapid rate at which technology is progressing, 2025 should actually see a quantum jump. Fully NCW-capable forces would be operational. Better Precision Guided Munitions (PGMs) would be available, including high energy lasers, plasma, electromagnetic, ultrasonic and DEWs. Information, surveillance, reconnaissance (ISR) and communications would be revolutionised. Long-range strategic aerospace platforms would be in play. Stealth and smart technologies and artificial intelligence would be optimised. Improved nukes would be more compact and lethal. Nano weapons and equipment, including micro UAVs, ant robots and the like would be utilised. Cyber warriors, worms, viruses and cybugs would be common. ASATs would be in use. Considerable progress can also be expected in the ongoing research of mind control, albeit this too can have adverse effects for mankind, should it fall into the wrong hands. Most of these technologies, as discussed, would be in use in China by 2025 and some of them, by extension, in Pakistan.

Technological advancements will greatly affect the manner in which wars will be fought. Conflict will be five-dimensional to include aerospace, land, sea, electromagnetic and cyber. Information warfare will include NCW, C^4I^2 Warfare, Electronic Warfare, Cyber Warfare and all other forms of operationalised cyberspace. Space combat, cyberspace combat, radiation combat, robotic combat and nanotechnology combat will add to the forms of combat. Operations will be increasingly inter-agency, involving greater application of all elements of national power. States like Pakistan will continue to employ hi-tech irregular forces. Asymmetric wars will be an ongoing affair. Information superiority will be as important as superiority on land, sea and aerospace. At the national-level, constant synergy would be required. De-conflicting actions would be necessary to achieve a united national front. The ability of the military to present a joint front will prove to be an absolute imperative, and the ability to conduct integrated operations with other components in the security sector a necessity.

[11]The strategy that countries should adopt for their military resources continues to be debated in the US. In the context of the war in Vietnam and, more recently, in situations of insurgency in Iraq and Afghanistan, proponents of Fourth Generation Warfare (4GW) continue to argue for less emphasis on the use of Information Technology (IT) in warfare. Meanwhile, US President Barack Obama's military strategy contains elements of the 4GW approach and also those of NCW and the exploitation of information technology. NCW continues to attract support as the approach that modern military forces should adopt. The controversy over drones appears to have intensified as more are used in military operations and their use in non-military roles increases. Debates over the impact that this will have on society continue. Cyber terror is also gaining increasing attention, with some people arguing that it is a serious threat, while others state that it is unlikely to have a significant impact. The relationship between war and technology and whether or not technology has altered the 'technology of war' is being debated. [12]It is surmised that 'technology of war' will always remain superior to technology per se. Today's commanders, both senior and junior, armed with modern technology and with the

---

[11]Roland, Alex, 'War and Technology', Foreign Policy Research Institute, Vol 14, No 2, February 2009, http://www.fpri.org/footnotes/1402.200902.roland.wartechnology.html

[12]Hoffman, Frank G, 'Hybrid Threats: Reconceptualizing the Evolving Character of Modern Conflict', http://www.ndu.edu/inss/docuploaded/StrategicForum_240.pdf

means and skills to exploit these resources, stand a better chance to be effective than great captains of yore such as Alexander the Great and Sun Tzu. If this had not been true, the concept and mechanics of warfare would have remained fixed in time but thanks to technological advancement, the whole concept of warfare has come a long way since the days of Sun Tzu and Alexander. Even for the veterans of WWII, it would be mind-boggling to visualise the panoply of the modern hi-tech battlefield strewn with a complex network of airplanes, missiles, tanks, drones, satellites, computers, GPS and an array of hi-tech command and control network which they may never have dreamed about.

Sub-conventional Combat

A broad division of the conflict spectrum will, in descending order, encompass nuclear conflicts, conventional conflicts, sub-conventional conflicts and conflicts in cyberspace. The sub-conventional has emerged as the favourite in recent years, as irregular forces have demonstrated greater strategic value than the conventional and nuclear forces. [13]Even the US and National Atlantic Treaty Organisation (NATO) forces have battled irregular forces and are themselves now engaging in hybrid wars by optimising proxy radical forces available on hire, as evident in Iraq, Libya and Syria. All the three Services of the Indian military will be engaged throughout the spectrum of conflict, the Navy even battling the sub-conventional at sea and their installations and establishments on land. Moreover, states both big and small will continue to use proxies in pursuit of their national objectives. The thinking that sub-conventional warfare and irregular forces are only used by a country against a conventionally superior adversary is passé. Militarily strong countries, such as the US and China,

---

[13]Katoch, PC, 'Special Operations', United Services Institution of India, New Delhi, 3 June 2013, file:///C:/Users/tara/Documents/All%20Articles%20Published/USI%20-%20 Special%20Operations.htm

employ sub-conventional means in terms of irregular forces and proxies. China has spawned Maoist insurgencies the world over, engaged with the Taliban and al-Qaeda and armed and supported Maoists in India and Nepal, while the US's support for the Wa Army in Myanmar are proof of this. Technology has also empowered terrorists. Many irregular forces, mostly state-sponsored, have acquired capabilities comparable to those of conventional forces. These irregular forces are making good use of IT and are more than familiar with modern communication. The advent of weapons of mass disruption is fairly advanced and the likelihood of weapons of mass destruction, as part of sub-conventional warfare, appears not too far away.

<h1 style="text-align:center">2</h1>

Threats and Solutions 2025

On the threshold of being the third largest economy in the world, India will surpass China's population by 2028, as assessed by the UN World Populations Prospect Report 2010. By then, we will also be one of the world's most connected and IT-savvy societies. Our superpower status, however, will depend on how best we manage the evolutionary process, economic growth, social change, resources and leapfrog technology in areas we lag behind. Advancements in technology have revolutionised warfare but we also see the [1]advent of 'smart wars' and 'no contact wars', coupled with sub-conventional wars and optimal employment of irregular forces within the matrix of asymmetric war. [2]Options of conventional conflict continue but are reducing because of multiple factors like prohibitive costs and lethality of weapons capable of inflicting enormous casualties. With continuing volatility in our neighbourhood, India faces the

---

[1]Klonoski, Jake, 'Fighting Smart wars in Smart Ways', *Oregonlive*, 23 August 2011, http://www.oregonlive.com/opinion/index.ssf/2011/08/fighting_smart_wars_in_smart_w.html

[2]Tertrais, Bruno, 'The Demise of Ares: The End of War as We Know It', *The Washington Quarterly*, Summer 2012, http://dx.doi.org/10.1080/0163660X.2012.703521

unmistakable collusive China–Pakistan threat, both in the conventional (in nuclear backdrop) as well as asymmetric domain. This is likely to be coupled with activation of space and cyberspace that potentially can spill over to any segment of the conflict spectrum including hi-tech war. In such a scenario, there is an urgent need for holistic appraisal of the future battlefield and the measures India must take.

Changed Face of Conflict

The dimensions of modern conflict and the impact of technology on warfare have been discussed in Chapter 1, highlighting the fact that operations will increasingly require inter-agency response and involve greater application of all elements of national power. What needs reiteration is the volatility in the sub-conventional segment of the conflict scenario, the activation of which is being increased by both weaker and powerful nations. [3]There is every possibility that countries will continue to employ hi-tech irregular forces. [4]The US launched its Global War on Terrorism to eliminate al-Qaeda but is alleged to have subsequently used al-Qaeda in Libya and in Syria, the latter including Pakistani Taliban. Similarly, China is believed to have armed Shia rebels in Iraq, trained the Taliban even as the US was invading Afghanistan, and later provided arms and advisors to fight US/NATO forces in Afghanistan. Bill Gertz and Rowan Scarborough, in an article titled, 'Inside the Ring—China-trained Taliban', quoted US intelligence in the *Washington Times* dated 21 June 2002 and said that China was training Taliban even before the 9/11 terrorist attack in the US. On 20 December 2010, in an article titled 'China's Military Aid to Taliban', Joshua Kucera reported a British military officer telling the *Aviation Week* that China may

---

[3]Bowden, Mark, *Black Hawk Down: A Story of Modern War*, Penguin Books, London, 2000.

[4]Nimmo, Kurt, 'The Real Benghazi Story: US Op to Arm al-Qaeda in Syria', infowars.com, 08 May 2013, http://www.infowars.com/the-real-benghazi-story-us-op-to-arm-al-Qaeda-in-syria/

have provided advisors to Taliban battling NATO forces and that Chinese specialists have been seen training Taliban fighters in the use of infrared-guided surface-to-air missiles. [5]In Myanmar, China is said to have established a deadly proxy in the United Wa State Army (UWSA) by supplying them with assault rifles, machine guns, rocket launchers, shoulder fired air defence missiles, armoured vehicles and helicopters armed with air to air missiles, as reported extensively by the international media. [6]In India, China, along with Pakistan, is supplying sophisticated weapons and communication equipment to Indian Maoists and has even given them arms manufacturing capability.

The Threats

The range of threats in the twenty-first century have become varied and highly complex. [7]There is a growing number of independent international and transnational actors playing power games on multiple levels, revolving around national, regional and global dynamics. [8]Current threats are less predictable than traditional state-centric ones and come from more diverse sources. [9]The level of uncertainty in the world has increased significantly. Computer hackers and criminals, disaffected domestic groups, natural and man-made viral borne illnesses and radical terrorists are all representatives

[5]Naing, Saw Yan, 'China Sells Helicopter Gunships to UWSA: Report', *The Irrawaddy*, 30 April 2013, http://www.irrawaddy.org/archives/33350

[6]Katoch, PC, 'Surreal China—Lessons Nextdoor', Centre for Land warfare Studies, New Delhi, India, 07 June 2013, http://www.claws.in/Surreal-China---Lessons-Nextdoor-Prakash-Katoch.html

[7]Gray, Collin S, 'Security Threats in the 21st Century', University of Reading, UK, November 2006, http://www.dtic.mil/cgi-bin/GetTRDoc?AD=ADA481210

[8]Thorton, Rod, *Asymmetric Warfare: Threat and Response in the Twenty-First Century*, Polity Press, USA, 2007.

[9]Patrick, Charles, 'India: Future of Change-Will the Greatest Future Security Threats to India be from Inside or Outside the Country?', UK, 2011, http://www.indiafutureofchange.com/featureessay_d0091.htm

of these new threats. Due to globalisation, more nations and non-state actors than ever before are active on the international-level. Regional issues have proliferated and often appear to threaten wider international peace and security. Terrorist groups and other non-state actors have taken advantage of regional conflicts and insecurities. India faces threats from a variety of sources such as nuclear missiles, cyber and cross-border terrorism, infiltration, demographic assault, conventional conflicts and insurgencies. In addition, India is battling asymmetric wars waged by China and Pakistan.

[10]India faces threats to its security from almost all its neighbours, be it the spillover of their domestic ethnic conflicts, large-scale illegal migration or providing a base for terrorism directed against India. [11]The Ministry of Home Affairs (MHA) has banned thirty-five terrorist organisations operating in India, but many more like Popular Front of India (PFI), that picked up arms against the Indian state in 2010 have not still been outlawed. Similarly, the National Army of Meghalaya, that the security forces have been battling, has not been banned either.

Taliban's hold over Afghanistan post 2014 and Chinese presence in Pakistan occupied Kashmir (PoK), South Asia and Indian Ocean Region (IOR) will further enhance collusive threat. Increased influence of Taliban in Afghanistan post 2014 will provide Pakistan her cherished 'strategic depth', best explained by [12]Robert Kaplan in his book, *The Revenge of Geography*. He wrote, 'An Afghanistan that falls to Taliban sway threatens to create a succession of radicalised Islamic societies from the Indian-Pakistani border to Central Asia.

---

[10]Dubey, Muchkund, 'India facing threats from almost all its neighbours', *Indian Express*, 02 September 2012, http://www.indianexpress.com/news/india-facing-threats-from-almost-all-its-neighbours-says-former-indian-foreign-secy/996696/

[11]Banned Terrorist Organizations, Schedule I - First Schedule (of the UA(P) Act, 1967), http://www.nia.gov.in/banned_org.aspx

[12]Kaplan, Robert D, *The Revenge of Geography*, Random House, New York, USA, 2012.

This would be, in effect, a greater Pakistan, giving Pakistan's Inter-Services Intelligence (ISI) the ability to create a clandestine empire composed of the likes of Jallaluddin Haqqani, Gulbuddin Hekmatyar and the Lashkar-e-Taiba: able to confront India in the manner that Hezbollah and Hamas confront Israel'.

[13]India faces collusive China–Pakistan asymmetric war with overlaps of conventional conflict under the nuclear shadow. The two-and-half front threat is real, with the possibility of the half front expanding. Insurgencies in India are likely to continue unless a social change can be managed. The declaration made in 2012 by Maoists to infiltrate personal security of political leaders in states and the Centre is an indication of the escalation of their activities. The massacre of Congress leaders on 25 March 2013 and issue of a fresh target list which had names of leading Congress leaders at the Centre, is further proof of how serious the Maoists are about their mission. In all probability, the Maoist insurgency will be upgraded with willing Chinese and Pakistani support under the cover of brazen denial. Introduction of shoulder-fired air defence missiles and use of uranium-based Improvised Explosive Devices (IEDs) may well be the next step. Cross-border terror may escalate. LeT – al-Qaeda footprints in Maldives, Kerala and efforts to link with the Maoists bodes more danger, particularly for South India. The current opposition party in Bangladesh is known for its links with anti-India terrorist organisations and so, a change of guard in Bangladesh could increase our problems.

Sun Tzu is believed to have said, 'The essence of warfare is creating ambiguity in the perception of the enemy'. China will in all likelihood upgrade sub-conventional pressure to include more

---

[13]Singh, Mandeep, 'Asymmetric Wars in the Indian Context', Institute of Defence Sudies and Analyses, 13 October 2011, http://www.idsa.in/idsacomments/Asymmetric WarsintheIndianContext_msingh_131011

intrusions along the border, thus increasing the dilemma of Indian leadership. China considers 'mind of the enemy' leadership the centre of gravity. This they already appear to be doing successfully as was evident during the 19 kilometre intrusion in the Depsang plains of Ladakh (April 2013). We may draw solace from the fact that we managed to coax them to remove their tents but actually China was a clear winner—sitting in Indian territory for full twenty-five days; displaying a 30-feet banner (televised globally), saying it is Chinese territory and making India demolish construction in Chumar which they may use later as justification for a diplomatic success. Some US scholars opine that China may even use tactical nuclear weapons to subdue India into territorial submission. The proliferation of Pakistan's tactical nuclear weapons (ostensibly in response to India's Cold Start doctrine) also needs to be viewed in context of Pakistan using tactical nuke(s) on pincer(s) of our Strike Corps deep inside the country. Would that justify a nuclear riposte on Pakistan and in what measure? We need to review our nuclear policy in this context. Ironically, tactical nukes were being discussed in military courses in our army more than two decades ago but it appears to have been stopped altogether.

India must be prepared for full activation of space, cyberspace and electromagnetic domains, taking into account China's credible capabilities. Already, China has continued satellite surveillance of our border areas and has demonstrated anti-satellite kill capability. In all probability, China has mounted weapons in space and as in the case of the US and Russia, is likely to develop laser and plasma weapons. China's cyber warfare capabilities are perhaps at par with that of the US. Cyber warfare may emerge as more dangerous than even nuclear threat since it has the potential to cripple critical infrastructure and the capacity to jam military communications, weapon systems and even misdirect missiles, given the ambiguity of origin of attack.

Significantly, China has been training some 600 People's Liberation Army (PLA) personnel in electromagnetic warfare annually.

Military Requirements in 2025

[14]The only viable approach to national security in the twenty-first century is to maintain an adequately sized, trained and equipped force that is capable of dissuading, deterring and when required, defeating a diverse set of adversaries. Swami Vivekananda had said, 'We are responsible for what we are, and whatever we wish ourselves to be, we have the power to make ourselves. If what we are now has been the result of our own past actions, it certainly follows that whatever we wish to be in future can be produced by our present actions; so we have to know how to act.' There is plenty of merit in what he said. We may not be able to catch up with China economically and in military terms, but we must aim to bridge the asymmetry in the sub-conventional, cyber, space and electromagnetic including optimising strategic partnerships and developing new ones like Taiwan has done. Defence budgetary allocations may need sustained boost till such asymmetry is removed, especially since insurgencies preclude reducing boots on the ground. [15]Multiple requirements that need to be met are: Networked elements of national power; information dominance and information assurance; ability to paralyse enemy C^4I^2 infrastructure; credible deterrence against state-sponsored terrorism; long-range expeditionary strategic forces; stand-off weapons to pre-empt enemy attack; mix of DEWs, PGMs, ASATs weapons etc; ability to disrupt enemy logistics/sustenance; mix of hard kill and

---

[14]Winter, Donald C, 'Adapting to the Threat Dynamics of the 21st Century', *The Heritage Foundation*, 15 September 2011, http://www.heritage.org/research/reports/2011/09/adapting-to-the-threat-dynamics-of-the-21st-century

[15]'Emerging Threats in the 21st Century-Strategic Foresight and Warning', Center for Security Studies (CSS), ETH Zurich, December 2007, http://www.isn.ethz.ch/Digital-Library/Publications/Detail/?lng=en&id=47160

soft kill options; layered strategic air and theatre missile defence; competitive cyber warfare capability; ability to exploit space and cyberspace and conventional forces capable of winning hi-tech wars.

We must invest in all domains of Diplomatic, Information Operations, Military and Economic (DIME) in areas of our strategic interest, especially our immediate neighbourhood. The third century Indian politician Kautilya had said, 'The arrow shot by an archer may or may not kill a single person; but skilful intrigue, devised by wise men, may kill even those who are in the womb.' China appears to be following this advice but not India. Internally, de-conflicting actions are required to achieve a united national front. Military jointness is an absolute imperative in the armed forces. As far back as in 2004, Prime Minister Manmohan Singh had said, 'Reforms within the armed forces also involve recognition of the fact that our Navy, Air Force and Army can no longer function in compartments with exclusive chains of command and single service operational plans.' Sub-conventional threats require countering by the security sector, of which security forces are one part. Ability to conduct integrated operations with other components of security sector is necessary. The entire citizenry should form a part of intelligence operations in the matrix of 'billion eyes on the ground' and for de-radicalisation programmes.

Both China and Pakistan possess advanced sub-conventional capability but India is lagging behind. This strategic asymmetry needs to be addressed on priority. We have not employed our Special Forces to create deterrence. Special Forces are ideally suited to control fault-lines of the adversaries without any signature or with ambiguous signatures and for shaping the environment in areas of our strategic interests. They do not create resistance movements but advice, train and assist resistance movements already in existence. Special Forces provide a range of strategic options to the political authority and must be employed strategically in sync with national security objectives. [16]There is an urgent need to develop publicised overt capabilities and deniable covert capabilities as deterrence against irregular war that may be thrust upon. We must have the will to selectively demonstrate this capability to ensure its credibility.

Similarly, our cyber capability should aim to stop enemy from: Accessing/using our critical information, systems and services; stealthily extracting information from enemy networks and computers including vulnerabilities, plans and programmes of cyberattack/war; penetrating enemy networks undetected and stealthily inserting dormant codes; manipulating and doctoring radio transmissions; destroying enemy computer networks; and if necessary develop and manipulate perceptions of adversaries. [17]The ultimate response to nukes is space-based lasers. We are still at the level of Laser Dazzler for crowd control. The DRDO is developing a vehicle mounted laser-based DEW system (Aditya) as technology demonstrator and a 25 kW laser system for hitting a missile in terminal phase at 57 kilometre. We must focus on leapfrogging to space-based lasers to be developed through strategic partnerships.

[18]A vital shortcoming is a lack of impetus in Revolution in Military Affairs (RMA), considering we are yet to define a National Security Strategy, National Security Objectives and a strategy to fight asymmetric war. We even lack an integrated C^4I^2SR system. It is no secret that we lack institutionalised strategic forethought which is further compounded by the fact that we have an inadequate higher defence organisation. Military modernisation is not meaningful and

[16]Katoch, PC and Datta, Saikat, *India's Special Forces—History and Future of Indian Special Forces*, Vij Books India Pvt Ltd, New Delhi, India, 2013.

[17]'The Space-Based Laser Integrated Flight Experiment-Global Missile Defense in the Boost Phase', *Team SBL-IFX*, http://www.wslfweb.org/docs/SBLWP.pdf

[18]Sahgal, Arun and Anand, Vinod, 'Revolution in Military Affairs and Jointness', *Journal of Defence Studies*, Volume 1 No 1, Institute for Defence Studies and Analyses, http://idsa.in/system/files/jds_1_1_asahgal_vanand.pdf

we lack military jointness and interoperability. ISR and intelligence is poor and there is inadequate synergy within the security sector. Investments and focus on R&D is minimal what with DRDO and PSUs being commercialised and unfocused. The involvement of private industry in defence is full of avoidable hurdles. Specialisation is misunderstood both at the national and military-levels. There is inadequate focus on equipping forces with cutting edge technology despite bulk operations being conducted at sub-conventional levels and the civil and military bureaucracies competing with each other in causing delays and creating hurdles.

The initiatives required by the government should include an Act of Parliament to usher in the much-needed RMA and to include nomination of the political authority to lead RMA. Ministry of Defence needs to respect and improve its response to proposals by the Services. Presence of politico-bureaucratic inertia requires the military to redouble efforts and galvanise military will, in turn building political will. Building public awareness is essential to mould national opinion and decision-making. The military must define a doctrine for NCW without further delay. The government should appoint a Chief of Defence Staff, making him a permanent member of the National Security Council (NSC) and Cabinet Committee on Security (CCS), merge IDS with MoD, interfaced with Ministry of External Affairs (MEA) and MHA, institutionalise strategic thinking within MoD, create Integrated Special Forces Command and a National Cyber Command and post both serving and veteran military officers on staff of a full time National Security Advisory Board (NSAB). The military should push for Integrated Theatre Commands (ITCs)—five to six based on geographical regions in addition to the Andaman & Nicobar Command (ANC) and Integrated Functional Commands (IFCs)—aerospace, cyber, air defence, two to three training, two to three Logistics and Maintenance Commands in addition to the Strategic Forces Command (SFC). In the past there has been talk of Expeditionary Strategic Forces like an Air Assault Division and a separate Marine Corps. The immediate emphasis should be on raising of the Marine Brigade (a case for which is languishing with the MoD for over a decade now) and allocating it to ANC, enhancing lift capability of Parachute Brigade, re-raising of erstwhile second Parachute Brigade and consolidation of Special Forces. In the overall jointmanship paradigm of the military, true jointmanship exists mainly within HQ IDS. The balance is thoroughly patchy. We need to move from 'de-conflicted operations' to 'joint operations' and eventually get to the stage of 'integrated operations'. Military must push for integration at operational and tactical-levels, vertically and horizontally, and institutionalise periodic review of jointness and interoperability.

Another vital step required for the military is to accelerate establishment of an integrated C^4I^2SR system. Presently, the Services do not even have common data structures, symbology and interoperable protocols. A true 'System of Systems' approach has yet to come through. Integrated communications must be established to provide seamless connectivity, both vertically and horizontally. All platforms must be network enabled. Cybersecurity must graduate to information assurance and information dominance. A holistic review should be done to ascertain requirements of stand-off PGMs, DEWs and ASATs. Technologies like Steerable Beam Technology, Wide Band/Software Defined Radios, Network Security, Common GIS, Data Fusion and Analysis, alternatives to GPS, Dynamic Bandwidth Management, lasers to shoot UAVs, Camouflage and Concealment etc. should be exploited. Simultaneously, we should not lose focus on equipping the soldier with the cutting edge. This is even more important to the fighting elements of the security sector (military, para-military, central armed police and police units) engaged in counterterrorist and counter-insurgency tasks. Government must focus on indigenous production of critical hardware, software, telecom equipment and chip manufacture.

The Defence Procurement Procedure 2013 is a good opening for the private sector but it still does not meet the requirements of information and communication systems as the procedures are too elaborate. By the time they are brought into operation, the technology will get outdated. The delays are primarily because of inadequate capacity of DRDO/PSUs who at times, in order to buy time, stonewall private sector participation. In order to stop all this, it would be better if all Information and Communication Technology (ICT) intensive projects are made 'Make' projects with full private sector participation. There is also a crying need to simplify and shorten the 'Make Procedure'. Additionally, both the civil and military leadership should lend themselves to attitudinal change to accommodate the concept of NCW, thereby adapting to the changing nature of war.

The task of the intelligence agencies needs serious review. If we get to know from [19] *The New York Times* that 11,000 Chinese are working on fourteen projects in PoK, then there is something drastically wrong with our information system. These intelligence agencies (most without a formal charter) need to be brought under parliamentary oversight and with proper legal structures. Human Intelligence (HUMINT) requires revival and the Defence Intelligence Agency (DIA) should be permitted to undertake its authorised mandate of operating transborder sources. India must employ its Special Forces strategically to shape the environment in its own favour and for surveillance of areas of our strategic interests. The Multi Agency Centre (MAC) should be expanded into the

---

[19] 'China deploys 11,000 troops in Gilgit area in Pakistan occupied Kashmir', *Daily News Analysis*, 28 August 2010, http://www.dnaindia.com/india/1430066/report-china-deploys-11000-troops-in-gilgit-area-in-pakistan-occupied-kashmir

National Counterterrorism Centre (NCTC) and connected to all concerned through the National Intelligence Grid (NATGRID), including the State Counter Terrorism Centres (SCTCs). National concurrence should be possible if the overall internal threat and a holistic plan to deal with it is discussed in totality rather than isolated discussions on NCTC, hostage rescue policy and the like. Technology should be exploited for real time dissemination of the Common Operational Picture (COP) and incorporation of a Decision Support System (DSS) to assist decision-making. The military should undertake holistic examination of its ISR and intelligence requirements. Networking of Services intelligence with 'all sources' intelligence and real time/near real time dissemination should be ensured.

Government must identify focus areas for R&D. There is an urgent need to leapfrog technology to bridge the gap vis-à-vis China. There is a positive need to slash the business empires of the DRDO and focus them on critical areas. We must open the private industry to the defence sector. A ruthless follow-up of DPP 2013 is needed. DRDO and PSUs must be made accountable and responsible. Groups of private industry should be identified for focused development. R&D allocations should be reviewed and appropriate share must go to private industry/group(s) of private industry tasked for defence requirements. DRDO/and PSUs must practice reverse engineering to catch up with the latest technology.

Information revolution and networked environment give rise to various entities. Services must retain core competency and have the ability to integrate with other domain specialists. Concept Development Centres (CDCs) should be established to involve modelling, simulation and synthetic environment that will provide a powerful aid to visualisation analysis, test, evaluation, training and rehearsal throughout acquisition life cycle. Simulation is the best way to understand and optimise dynamics of manufacturing process and support chains. CDCs require networking with knowledge entities. We must establish a National War Gaming Centre.

Technological advancements will make future battlefields complex. By all indications we are likely to continue to face threats, particularly from China and Pakistan, in the sub-conventional, space, cyberspace and electromagnetic. Our nuclear triad is getting operational but the widening military asymmetry between the PLA and the Indian military must be bridged by enhancing our capabilities in the sub-conventional, space, cyberspace and electromagnetic fields. We urgently need an RMA to take us into the next level of military potential and attain a national synergy to meet future challenges. We can hardly afford to be complacent.

3

Revolution in Military Affairs

Revolution in Military Affairs[1] can be considered a phenomenon that is nearly three decades old. Soviet military thinkers first dabbled in RMA (though the term was not coined by them) during the period 1960-70. The Soviet experiment was primarily with respect to the impact of nuclear weapons and Inter-Continental Ballistic Missiles (ICBMs). Their focus was to dovetail the employment of nuclear weapons into their warfighting doctrine, giving them the cutting edge in future wars. More than a decade later, in the mid-eighties, the Chief of Soviet General Staff, Marshall Nikolai Ogarkov, revived the debate over RMA with reference to precision-guided conventional weapons. [2]The concept caught the fancy of the US much later, who actually coined the term RMA. Since then militaries worldwide have been experimenting and adopting RMA.

---

[1] Chapman, Gary, 'An Introduction to the Revolution in Military Affairs', 2003, http://www.lincei.it/rapporti/amaldi/papers/XV-Chapman.pdf

[2] The United States Army 1995 Modernization Plan. Force 21, 6 April 1995, http://oai.dtic.mil/oai/oai?verb=getRecord&metadataPrefix=html&identifier=ADA286744

Chinese interest in RMA and the structure of future US armed forces is strong to the extent that RMA is now being incorporated into the Chinese strategic military doctrine. Their interest in the RMA theory and practice was accelerated due to the dramatic and speedy US victory over Iraq in the 1991 Gulf War wherein US dominance was achieved through precision weaponry, satellites and superior ICT. The power of technological advances coupled with matching strategy and concepts, organisation and training was fully apparent. This was a catalyst for the PLA to get going on the path to 'informisation'.

Definition

Throughout history, advances in technology and strategy have revolutionised the way wars are fought. Many definitions have been coined to describe the nuances of RMA. [3]Information in the public domain says that, 'The military concept of RMA is a theory about the future of warfare, often connected to technological and organisational recommendations for change.' [4]RMA results when a nation seizes an opportunity to transform its military doctrine, training, organisation, equipment, tactics, operations and strategy in a coherent pattern in order to wage a war in a novel and more effective manner. There are many definitions of RMA but in simple terms it can be defined as, 'A major change in the nature of warfare brought about by the innovative application of technologies, which combined with dramatic changes in military doctrine and operational concepts, fundamentally alters the character and conduct of operations.' Transformation is essential to cope with these changes and most countries have put in place organisations dedicated to conceptualising and implementing transformation.

---

[3]Joint Services Command and Staff College Library, *Research Guide Series, Revolution in Military Affairs (RMA)*, http://www.da.mod.uk/colleges/jscsc/jscsc-library/.../research.../rmainternet.pdf

[4]Gray, Collin, *Strategy for Chaos—Revolution in Military Affairs and the Evidence of History*, Crown House, London, 2005.

Perspectives

Global debate[5] on RMA is centred on the following perspectives:

- The first perspective highlights the political, social and economic factors worldwide which might require a completely different type of military and organisational structure to apply force. It focuses primarily on changes in the nation state and the role of an organised military in using force.

- The most common 'System of Systems' perspective on RMA highlights the evolution of weapon and information technology, and military organisations and doctrine. These include three overlapping areas for force assets: Information, Surveillance and Reconnaissance (ISR), Command, Control, Communications and Intelligence (C^3I) and Precision Force.

- The third portrays the pessimistic view that a 'true' RMA has not yet occurred or is unlikely to occur since much of the technology and weapon systems ascribed to contemporary RMA have been under development for quite some time. However, the bottomline is that RMA is an ongoing phenomenon with no specific start or end point. Perhaps, it is akin to the pursuit of excellence which is a never-ending process.

Political Direction

Globally, militaries have researched and considered RMA as an organisational concept. [6]Those that have capitalised on it have had very strong and focused political direction and legislation to enforce systemised programmes for implementation of RMA.

---

[5]Ibrugger, Lothar, 'The Revolution in Military Affairs', Special Report, NATO Parliamentary Committee, November 1998, http://www.iwar.org.uk/rma/resources/nato/ar299stc-e.html

[6] Bernard, Simon, 'The Revolution in Military Affairs: Approach with Caution', *The Army Doctrine and Training Bulletin*, Volume 3, No 4/Volume 4, No 1, Winter 2000/Spring 2001, USA, http://www.army.forces.gc.ca/caj/documents/vol_03/iss_4/CAJ_vol3.4_13_e.pdf

Axiomatically, the infrastructure and investment demands are heavy and many countries have not invested required sums in defence, especially where the potential of RMA is not grasped by the establishment. A successful revolution also requires key bureaucracies to possess certain institutional characteristics that enable them to direct technological advances to dramatically improve military efficiency and efficacy. Developed countries that have had exponential increase in RMA have adopted a top-down approach emanating from the political apex. They made organisational changes, necessary to accelerate synergy in the armed forces despite critics pointing out that a 'revolution' within the military ranks might carry detrimental consequences, produce severe economic strain and ultimately prove counterproductive.

The Indian military requires organisational changes that are necessary to give an impetus to synergising the armed forces towards integration and achieving RMA. These changes have to be driven from the top political leadership of the country. [7]In the US, the catalyst for the transformation process commenced with former Secretary of Defence, Donald H Rumsfeld; the US Department of Defence created US Joint Forces Command (JFC) as the transformation laboratory of the US military to force the US armed forces into jointness. [8]The Goldwater Nichols Department of Defense Reorganisation Act of 1986 brought about revolutionary changes in the US armed forces, accelerating synergy and boosting RMA. [9]In China, change was ushered in by former President Jiang

---

[7]Watts, Barry D, 'The Maturing Revolution in Military Affairs (Pt 2)', Center for Strategic and Budgetary Assessments (CSBA), http://www.isn.ethz.ch/Digital-Library/Articles/Detail/?lng=en&id=162686

[8]Goldwater Nichols Department of Defense Reorganization Act of 1986, National Defense University Library, USA, http://www.ndu.edu/library/goldnich/goldnich.html

[9]Wu, Jun; Xiangli, Sun and Side, Hu, 'The Impact of Revolution in Military Affairs on China's Defense Policy', Institute of Applied Physics and Computational Mathematics, Beijing, China, 2003, http://www.lincei.it/rapporti/amaldi/papers/XV-WuRMAImpact.pdf

Zemin and its implementation overseen by the Central Military Commission and the Chief of General Staff of the PLA. In Germany, the transformation process was initiated by the Berlin Decree which aimed to integrate the armed forces, ensuring to reap full benefits of ongoing technological advancements. [10]The German Chief of the Defence Forces was appointed to oversee the transformation of the armed forces. In India, even a Chief of Defence Staff, though recommended by the Kargil Review Committee, is yet to be appointed. Jointness in Services simply has to be forced from the top. Perhaps, only an Act of Parliament can enforce this.

Technology and RMA

It is generally opined that RMA is driven by technological advancements made in the recent past in information technology and changes in the fields of communication, computer and network. This is only partly true; limiting RMA to only systems would be highly incorrect. Recent technological advancements actually require revolutionary changes in the manner in which we conduct our military business because RMA encompasses the entire military organisation.

Tenets of RMA

[11]There are four basic tenets: First, RMA is not simply technological but concerns significant progress and change in at least the important military-related areas of technology, organisation, doctrine and operational concepts. Second, changes or progress in the above areas of technology, organisation, doctrine and operational concepts by themselves do not represent a true RMA, rather it is the synergistic combination of these developments which forms a true RMA and

---

[10]Christian, OG, 'Understanding the Prussian-German General Staff System', 1992, http://www.dtic.mil/dtic/tr/fulltext/u2/a249255.pdf

[11]Mazarr, Michael J, 'The Revolution in Military Affairs: A framework for Defense Planning', Institute for Strategic Studies, USA, 10 June 1994, http://www.strategicstudiesinstitute.army.mil/pdffiles/pub242.pdf

alters the nature of warfare. Third, RMA emerges from revolutionary changes of historic magnitude within the broader social, economic and political environment of national and global societies, which in turn offer the conditions for RMA to be recognised, appreciated, internalised and exploited. Fourth, the smooth and successful process of recognition, appreciation, internalisation and exploitation requires flexibility, acceptability, innovation and openness to change, particularly on the part of the military.

Paradigm Shift

RMA involves a paradigm shift in the nature and conduct of military operations which creates new core competencies in dimensions of warfare that render obsolete or irrelevant one or more core competencies of erstwhile dominant players. Blitzkrieg or lightning war, used in the first years of WWII by Germany, created new operational and tactical-level models for land warfare, rendering static lines of defence obsolete. During WWII, carrier warfare created new tactical and operational-level models for battle at sea, rendering large battleships obsolete. Introduction of ICBMs has created a new dimension of warfare, initiating a new core competency of long-range, accurate delivery of high yield nukes. Anti-satellite capability, space warfare and cyber warfare have ushered in yet more core competencies. Rise in RMA has also enforced antidotes in the shape of asymmetric and fourth generation wars, the application of which is countering the RMA of US military in Afghanistan and Iraq to a considerable extent.

Force Application and Transformation

[12]Emerging trends of warfare have greater emphasis on the sub-conventional. The Army would require to continue as manpower intensive, and so, has to be equipped and networked for such

---

[12]Nanavatty, RK, Lt Gen, 'Changing Nature of Conflict: Trends and Responses', *Manekshaw Paper No 18*, 2010, Centre for Land Warfare Studies, New Delhi, http://www.claws.in/administrator/uploaded_files/1267073084MP_18.pdf

operations with the internal security organs of the state. The Navy is similarly faced with low-intensity maritime threats, heightened after 26/11 Mumbai terror attack and require to be addressed on a different plane. The Air Force, in support role, will need its own force multipliers, interoperable with sister Services. While moving up to higher levels of the war continuum, even the Army will need to be more machine intensive. At the tri-Service/national-levels, the far end of war would deal with the application of ballistic missile, anti-ballistic missile, defence and nuclear weapons. Each zone of the war continuum requires transformation and synergy with different instruments of state power and a differential in the level of man-machine interface requirements. Inter-Service jointness simply has to be enforced as fait accompli. Force application throughout the war continuum must be at the locus of engagement with the enemy. It must be 'joint' and could be virtual; not necessarily physical but including diplomatic, economic, information operations, military and economic domains. These national-level applications must be in place 365 days and 24x7; even during 'peace' since the military continues to be engaged in low intensity operations.

Need for RMA

[13]The military needs RMA to create positive asymmetrical capabilities and comprehensive competitive edge over adversaries in addition to transforming its current perception and thinking. Technological advances enable precision delivery of enormous firepower of hundreds of megatonnes on a given target, thousands of kilometres away. Destruction of satellites in orbit is possible. The capability has increased phenomenally in 'mass' and 'range' but the 'time' factor is frighteningly compressed; battle is being viewed and decided in almost real time. This has been possible because real time communication links and advent of aggressive media coverage has made it easy to

---

[13]Kak, Kapil, 'Revolution in Military Affairs – An Appraisal', Institute for Defence Studies and Analyses, New Delhi, http://www.idsa-india.org/an-apr-01.html

view the battle from our bedrooms. Everyone from the man on the battlefield to the entire chain of command and control, right up to the chief political executive and the people are on the same real time grid. The implication of all this for the military is that decisions have to be taken with greater swiftness and efficacy and the entire consultation/decision-making process has to be minutely reviewed. This calls for radical changes in our organisational structures, work culture, warfighting capabilities, doctrines and operational concepts.

Enhancing RMA

Factors that are relevant to RMA are time, technological capabilities, percentage adaptability of technology, human resource, inertia of the organisation, net-centric culture and security vulnerabilities. An increased use of high technology, well-trained human resource and a net-centric culture, which refers to organisational policies and strategies, help growth in RMA. Conversely, non-adaptive technology that does not lend to innovation and upgradation reduces the capability of RMA. An increase in network-centric culture (largely a state of mind) causes exponential rise in RMA. Measures for enhancing RMA include development of long-range precision attack capability and integration of civilian hardware industry for defence production in common use technological areas and specific security measures, particularly in the telecommunication sector to ensure proprietary protocols and standards through which they would have full control over the networks.

[14]NCW is perhaps the most important component of RMA. The challenge is to do it within an organised framework and with full security. NCW allows us to move from an approach, based upon the massing of forces to one based upon the massing of effects. This allows us to reduce our battlespace footprint, which in turn

---

[14] Blash, Edmund C, 'Network-centric Warfare Requires A Closer Look', IWS—The Information Warfare Site, USA, May 2003, http://www.iwar.org.uk/rma/resources/ncw/ncw-forum.htm

reduces risk because we avoid presenting the enemy with attractive high value targets. Once empowered with knowledge derived from a shared awareness of the battlespace and a shared understanding of the commander's intent, then our forces can display initiative to meet the commander's intent and be more effective when operating autonomously. An increase in NCW capabilities will exponentially enhance RMA.

China's RMA Experiment

[15]China is actively promoting RMA with Chinese characteristics and making focused advancements in national defence and modernisation of the armed forces. Top Chinese military and civilian officials have periodically affirmed the importance and relevance of RMA to China's military modernisation. Inspired by the 1991 Gulf War and subsequent US actions in Kosovo, China is shaping modernisation of the PLA in the context of global trends in military transformation. To some Western analysts, Chinese RMA is limited to 'pockets of excellence' only since large amount of weaponry and technology is still imported. [16]It would be prudent for India to take note and monitor the Chinese RMA experiment, nuances of which are:

- Chinese military technologies and production capabilities are impressive. Massive technological progress has been realised in a short span, incorporating sophisticated foreign technology and enabling the PLA to take advantage of an RMA of leading foreign militaries. China has hitched its

---

[15]Ji, You, 'The Revolution in Military Affairs and the Evolution of China's Strategic Thinking', *Contemporary Southeast Asia*, Volume 21, 03 November, December 1999, http://www.jstor. org/discover/10.2307/25798464?uid=3738256&uid=2129&uid=2&uid=70&uid=4&sid=21 102365801387

[16]Guha, Manabrata, 'China and the RMA Today', *ISN*, Zurich, 11 April 2013, http://www. isn.ethz.ch/Digital-Library/Articles/Special-Feature/Detail/?lng=en&id=162697&contextid7 74=162697&contextid775=162694&tabid=1454241796

technological evolution to the global train by committing itself to a more open economy and breaking the paradox of the snail-paced 'self-reliance only' concept.

- In terms of field of force structuring, the Chinese have achieved growth in asymmetric capabilities. They have invested heavily in submarines and guided missile destroyers to counter a probable US Carrier Battle Group in the stand-off against Taiwan, making sea capability the answer to a superior US armed forces' sea control capability. The Chinese are not only investing in the PLA and military hardware but in all other domains such as diplomatic, information operations, military and economic including foreign governments in India's immediate neighbourhood and the Indian Ocean Region.

- PLA is being 'informised' and downsized, and the older leadership is being replaced with a younger and more technologically aware force. Thousands of all ranks are being put through advanced studies in RMA-related subjects on a yearly basis. Specific appointments for general officers are tenable by those who have had tri-Service experience or satisfy laid-down technological criteria.

- Capabilities being focused upon as part of RMA include:
  - Advanced ICBMs, nuke delivery systems and an undeclared chemical weapons capability
  - Advanced satellite and anti-satellite capabilities
  - Extensive third dimension capability, Rapid Reaction Forces, expanding Blue Water Navy and a formidable Air Force
  - Potent cyber warfare capability, political will to use it and increasing offensive potential in this context
  - Large chip manufacturing capability

- Ongoing organisational adjustments in addition to mere downsizing and reorganised logistics companies are being put in place
- Development and proactive employment of sub-conventional/asymmetric warfare capabilities

Indian Military

Recognising that RMA of the Indian military cannot be delinked from the Indian state, our current status is something like this:

- True synergy and jointness in the military is yet to be enforced by the political apex. Evolving a Joint Warfighting Doctrine by the Services cannot suffice in the absence of joint organisations that would execute such doctrine.
- Headquarters Integrated Defence Staff has come up as another Service HQ instead of being seamlessly integrated with the MoD. It lacks operational responsibility and authority.
- A Chief of Defence Staff is yet to be appointed despite specific recommendations made by the Kargil Review Committee a decade ago.
- A tri-Service NCW Doctrine and an Enterprise NCW Architecture are yet to be evolved.
- Tri-Service netcentricity is absent. Large number of command and control equipment and networks are being established but lack common standards and protocols. Several interoperability constraints within each Service exist.
- Lack of an Enterprise GIS deters exploitation of Geospatial Intelligence (GEOINT) and optimisation of information systems being deployed.
- Long-range stand-off precision attack capability against land targets is limited.

- Procurement of weapons platforms and equipment need to cater for technological adaptation; ability of technological entities to be integrated for synergised joint operations and the scope for innovation.
- Air defence capability is limited and differences between the Services with respect to control of air space remain unresolved.
- Establishment of the Tactical Communications System (TCS) is inordinately delayed.
- No dedicated defence band has been allocated in the spectrum. Adequate bandwidths are not available to exploit technology.
- Development of human resource is largely being done on individual Service basis. A study to determine tri-Service IT training is yet to take off.
- Cyber warfare capabilities in the military and at a national-level are at a nascent stage. Offensive capability is poor and no policy has been enunciated on this count.
- Policy for countering asymmetric/fourth generation war has not been evolved. National Information Grid is yet to be established. No national policy for employment of special forces exists.
- Negligible indigenous chip manufacturing capability.
- Enormous security vulnerabilities exist with practically all computer and communications equipment and large amount of software being imported. Chances of malware being embedded at the manufacturing stages are high.

Requirement

For a 'revolution', an 'evolutionary approach' will not suffice. At the national and military-level we need to be proactive. There is an urgent requirement to make a blueprint for RMA and legislate it through

Parliament as advanced countries have already done. The draft would perhaps need to be prepared by the military. Allocation of defence budget should then be in sync with the approved RMA blueprint. Surrender of unutilised portion of defence budget should be prohibited as in almost all cases, it is due to bureaucratic delays and unutilised funds get carried over to the next financial year. The government should appoint a CDS with full operational powers without further delay. The political hierarchy should make jointness in the Services a fait accompli—initiate the process for establishment of bi-Service/tri-Service commands by amalgamation of existing commands of the Services; unified command structures should be speedily ensured throughout the three Services horizontally and vertically in a time-bound manner, replacing existing arrangements; ensure commonality of equipment, particularly those that carry out common functions such as air defence, electronic combat, combat identification, application of precision force etc.

Steps should be taken to establish network centricity at the national as well as the military levels, taking a holistic view and adopting a top-down approach. The thrust towards developing comprehensive joint Services 'System of Systems' approach must be specific. A dedicated Defence Band from the spectrum is needed, given the security requirements. Establishment of a cyber command must be speeded, developing offensive capabilities and making information dominance an integral part of the cyber warfare doctrine. Comprehensive policies for countering asymmetric/fourth generation war and for employment of special forces need to be defined. Indigenous chip manufacturing capability needs to be established on a fast-track progressive basis. We must ensure self-reliance in software, hardware, production of hardened active network components including storage area networks and in the interim, develop testing capability for malware in imported software, hardware and communication equipment. Concerted efforts are required to achieve self-sufficiency in critical areas like space technology and development of modern weapon systems including comprehensive 'System of Systems'. In terms of training, education and joint training of military personnel to support and promote the new RMA with Indian characteristics is required. The defence procurement process must facilitate speedy 'System of Systems' approach.

RMA in the Indian military is ongoing but needs to be drastically focused and accelerated. The political hierarchy must enforce jointness in the military and ensure organisational changes that are necessary to give an impetus to synergising the armed forces into total integration. Advancements in military technology will enable operations to be conducted with such speed, precision and selective destruction that the manner in which future wars are fought and their consequent political impact will fundamentally alter the way in which military and political affairs are conducted. In order to respond to this imperative, as also to fully exploit the potential offered by these technologies, our organisation and processes, concerned with the whole business of producing military/security capability will need to be re-structured accordingly. We must be able to protect own information systems, attack/influence information system of adversaries and leverage own strengths to gain decisive advantage in a battlespace where national security is threatened. Future threats require us to leapfrog into total transformation through application of RMA.

4

Net-centric Warfare

A basic understanding of NCW comes from examining the three domains of conflict—physical, information and cognitive. There is also a fourth one, the social domain, which is now being talked about much more. The physical domain spans the traditional environments of land, sea, air and space in which conflict typically occurs. It is home to the platform and communication networks of a given military force. However, the physical domain provides an incomplete picture in capturing the complex interactions and outcomes of real warfare. This is the primary reason for including the information and cognitive domains in the conceptual framework of NCW. The information domain represents the realm in which information is created, manipulated and shared. Information traces its origins to data collected from sensing events in the physical domain. This domain also encompasses all the means of conveying decisions, plans and orders that translate a cognitive response into physical actions. Consequently, it is increasingly the information domain that must be protected and defended to enable a force to generate combat power in the face of offensive actions by an adversary. On the other hand, the cognitive domain is the locus of the functions of perceiving, making sense of a situation, assessing alternatives and deciding on a course of action. This domain exists within the mind of the warfighter and holds the intangible elements of knowledge, understanding, decision-making, morale and leadership.

[1]NCW is an information superiority-enabled concept of operations that generates increased combat power by networking sensors, decision-makers and shooters to achieve shared awareness, increased speed of command, higher tempo of operations, greater lethality, increased survivability and a degree of self-synchronisation. NCW relates to behaviour, both human and organisational. It is most important to develop a 'Network Culture' for application in military operations to ensure success. [2]NCW focuses on the combat power that can be generated from the effective linking of maximum warfighting entities. It is the ability of geographically dispersed forces to create a high-level of shared awareness that can be exploited for effective and efficient execution of operations to successfully achieve the intent of the commander. NCW is transparent to geography, mission and size of the force and has the potential to merge the tactical, operational and strategic-levels of military hierarchy, leading to the cohesive employment of disparate inter-Services resources. NCW is not technology alone but encompasses the gamut of emerging military response to the Information Age. The tenets of NCW comprise a robustly networked force that improves information sharing; information sharing in turn enhances the quality of information and shared situational awareness; shared situational

---

[1] Wilson, JR, 'Network Centric Warfare 21st Century', http://www.afcea.org.ar/publicaciones/wilson.htm

[2] Alberts, David S; Garsta, John J and Stein, Fredrick P, 'Net Centric Warfare: Developing and Leveraging Information Superiority', Department of Defense C4ISR Cooperative Research Program, USA, http://www.dodccrp.org/files/Alberts_NCW.pdf

awareness enables collaboration and self-synchronisation, and further enhances sustainability and speed of command, and finally, all of these dramatically increase mission effectiveness.

NCW operations exploit state-of-the-art science and technology to integrate widely dispersed human decision-makers, situational and targeting sensors, weapon platforms and field forces into a highly adaptive and comprehensive 'System of Systems' to achieve unprecedented mission effectiveness. For exploiting science and technology and to apply a technology-oriented framework, NCW needs a Surveillance Grid that rapidly generates battlespace awareness and self-synchronisation, an Information Grid that is a high performance network that provides the backplane for computing and communication, a Command Grid that principally is the province of human decision-makers but could include knowledge-based artificial intelligence and software applications that act as command advisers and are able to recommend courses of actions, and an Engagement Grid that exploits the awareness and translates it into increased combat potential. During the coming decades, the focus of the Indian Army (IA) will be on leveraging emerging technologies to integrate dispersed sensors, networks and weapon systems. [3]The thrust areas for technologies, and implications on acquiring NCW capabilities are sensors, networks, their integration and modern weapons. [4]The transformation requires change in our concept of operations, doctrines, organisations and force structure and above all, in the psyche of the soldier and the leadership. Associated changes in logistics, education and training will also be required. These changes will have to be concurrent and on existing structures so as

---

[3]Ahvenainen, Sakari, 'Backgrounds and Principles of Network-centric Warfare', National Defence College, USA, November 2003, http://www.carlisle.army.mil/dime/documents/ NCW%20Background%20Principles.pdf

[4]Oberoi, Vijay, Lt Gen, *Net-Centric Warfare, Centre for Land Warfare Studies*, KW Publishers Pvt Ltd, New Delhi, 2005.

to bring about a graduated increment in NCW capability within the constraints of development and implementation time.

Technology provides numerous options to build automated Command, Control, Communications, Computers, Information and Intelligence (C^4I^2) systems for effective leadership in the modern battlefield. Success in combat depends greatly upon fused, tailored intelligence which is communicated securely and rapidly. Speed is a crucial component. The critical elements of Sensor and Engagement Grids are hosted by a high quality information backplane. These are supported by value-adding command and control processes, many of which need to be automated to achieve speed. This, in essence, personifies the essential characteristics of a C^4I^2 system. Therefore, there is a need for a tremendously flexible and robust C^4I^2 architecture which functions as a process of organisations, doctrines and technologies. The Indian Army's doctrine for IW unambiguously recognises the increased role of IW due to exceptional growth in technology in the past few decades. IW, therefore, attempts to influence the first three activities of the Observe, Orient, Decide and Act (OODA) loop by disrupting enemy's observation and surveillance systems, corrupting enemy's orientation and misguiding his perception and finally, inducing him to arrive at a wrong decision. [5]This aspect of IW has significant implications for any C^4I^2 architecture. Information integrity, information assurance and information superiority, therefore, need to be key factors while conceptualising any C^4I^2 system.

The digital age promises to revolutionise the battlefield decision-making process. Here, again, time is the decisive factor. There is likely to be an overload of information from a variety of sources and there will be tremendous pressure on the decision-maker to keep up with the pace at which decisions would be required to be taken. In the

---

[5]Anand, Vinod, 'Joint Development of Inter-Services Network and C^4I^2 Systems', Institute for Defence Studies and Analyses, October 2009, http://www.idsa-india.org/an-oct-00-9.html

process of this balancing act between requirement of speed and need for quality, there is a likelihood of the quality of decisions being affected. It is in this context that an efficient and reliable C^4I^2 system will become indispensable.

Netcentricity in the Indian armed forces including in the IA has mushroomed bottom upwards. Lack of an NCW philosophy/doctrine has resulted in an ambiguous NCW architecture, which has still not been defined. There are doctrines for C^4I^2 and IW but these two spheres are components of NCW and do not constitute it by themselves. NCW must also encompass policies, strategy, concepts, military organisations and adjustments. To transform the Indian armed forces into an NCW-capable force, we need an NCW philosophy/doctrine as the starting point. Concepts of individual Services should flow from a joint doctrine. This will facilitate development of coherent tri-Service networked architecture. Prominent reasons for this void include late establishment of HQ IDS, its non-seamless integration with MoD, its lack of operational responsibility/authority and absence of a CDS. Lack of a top-down approach has resulted in flawed netcentricity, which can be gauged from the following existing state:

- Presently, networks of the three Services do not talk to each other; respective intranets are not interoperable either. Neither voice and data networks, nor radio communications, are interoperable to the desired degree. Radio sets differ in their frequency bands, wave forms and secrecy algorithms.
- Each Service develops networks on its own and starts thinking of interoperability at a much later stage. Though subsequent convergence is very much feasible, it does become difficult since technologies differ, apart from avoidable additional costs and time that is required to achieve interoperability.
- Notwithstanding the fact that the Defence Communications Network (DCN) itself is some years away, very little is being

done to achieve a successful Services handshake through it as C^4I^2 is not part of the project. This implies that the Army, Navy, Air Force and HQ IDS will need to develop their own software, with attendant problems of integration. The DCN, therefore, will come up as a tri-Services super highway with no traffic till the software is developed and integrated.

- Common standards and protocols for the three Services have not been evolved. Finalising and adoption of standards and protocols, mutually compatible database structures, development/deployment of interfaces between systems using disparate platforms and commonality of hardware are challenges which need to be overcome. Bringing the standards and protocols of the three Services on the same plane is a gigantic task that can only be solved through outsourcing, given the-levels of expertise available within the Services. This process is way behind, save a few in-house studies that are underway.

- Standard Operating Procedures and formats have not been synchronised.

- Common data applications are yet to be conceived.

- There is absence of knowledge management.

- Following a comprehensive study (including sustained interaction with the three Services and HQ IDS) undertaken by Military Survey (which supplies paper/digitised maps to the three Services and paramilitary forces), the IA had issued a GIS policy and a project for establishment of an Enterprise GIS at the Request for Proposal (RFP) stage. However, the Military Survey was moved out from the Directorate General of Information Systems in the Army HQ and placed under the Directorate General of Military Operations. No further work on GIS is being done by Military Survey.

- No single unifying secrecy algorithm for the three Services has been developed and no work in this direction has commenced, though technological solutions exist in the country and can be engineered. No coherent policy in this regard has been worked out.
- Adequate bandwidth is not available to exploit technology in absence of a dedicated Defence Band.
- Lack of a single branch in HQ IDS that can oversee C^4I^2, IW, EW, space, information assurance cum cybersecurity, communications, information systems, joint organisations etc. The proposed establishment of an ICT branch will only meet this requirement partially. The requirement is of an all-encompassing 'Transformation Branch'. These components of RMA require to be viewed holistically under a single head in order to ensure a top-down revolutionary approach to NCW.

Within the military, an NCW doctrine has been under preparation for many years but is still not finalised. Lack of a top-down approach has adversely affected netcentricity at tri-Services level and within the Army with ambiguous division of responsibilities between the stakeholders of netcentricity (Operations, Perspective Planning, Signals, Information Systems), most of whom resist change and want to exert authority from within their comfort zones. The fact that the Tactical Command, Control, Communications and Information (TacC^{3}I) in essence is the mainstay of netcentricity in the IA is not recognised by some in authority. Inordinate delays in developing and deployment of the Tactical Communications System (TCS) as well as the interim TCS are severely restricting the test beds of the TacC^{3}I and eventual fielding of its subsystems like the Command Information Decision Support System (CIDSS), Battlefield Surveillance System (BSS), Air Defence Control & Reporting System (ADC&RS),

Battlefield Management System (BMS) etc. The only Operational Information System (OIS) fielded till now is the Artillery Command Control and Communication System (ACCCS). The policy on data handling and data storage is yet to be enunciated and is being handled by Signals instead of IS. Similarly, no policy on simulation and war-gaming has been enunciated. The requirement of a bandwidth in sync with increasing netcentricity is still being worked out.

In order to modernise, the Infantry intends to handle computer and radio subsystems and software integration of Project F-INSAS all by themselves instead of letting IS handle the same as part of the Battlefield Management System (BMS) programme, with the latter charged ab initio with Army-wide integration at regiment and battalion-level, including time and costs savings. The Army intranet has not been made fully secure because of which implementation of e-learning is severely hampered with most course materials classified. Army Wide Area Network (AWAN) has been fielded on the Army intranet but last mile security is still to come. Numerous policies on cybersecurity have been issued by different agencies instead of a single comprehensive one. Cybersecurity is under Signals instead of IS and there is continued resistance to the proposal for setting up an AISS under the latter. Development and fielding of the TacC^{3}I frequently hits roadblocks, which though temporary, do considerably delay these projects. In-Service problem areas are resistant to change, perceived as danger to individual turfs, and in some instances, there is lack of understanding of technology by two-and three-star level officers who hold decision-making appointments. In contrast, the Chinese are ensuring systematic institutionalised technological training before general level officers get posted to certain specific appointments. We need to follow suit. There is also a baseless belief that while netcentricity is good for conventional operations, it will not be of much use in anti-terrorism/fourth generation warfare type of operations.

The processing of cases get inordinately delayed because of lack of automation of directorates at Army Headquarters (proceeding at excruciatingly slow pace), none at MoD-level and no effort towards speeding this up and integrating them for faster processing of cases. Then there are problems of inter-Service egos. The Army intranet has not been extended to the Defence Services Staff College (DSSC) and College of Defence Management (CDM) on the plea that, being tri-Service organisations, they come under HQ IDS. This is despite the fact that over 90 per cent of students and staff in these organisations are from the Army and would benefit most from e-learning through Army intranet. Even HQ IDS is not extended to Army intranet on the same plea. Even the few Army Static Communications Network (ASCON) telephones with HQ IDS were withdrawn sometime in 2006-2007. So much for jointness.

To ensure effective transformation from the Platform-centric capabilities to the Network-centric, a phased shift in the existing technology at the level of all the three Services and horizontal fusion in our armed forces at laid-down hierarchical structure is necessary. A thorough cultural change is warranted through an aptitude-based selection procedure which should influence the values, attitudes and beliefs of our future military leaders. We need to pursue human resource development vigorously. Requisite emphasis needs to be laid on educating the man behind the weapon on the state of future warfare and how to fight it. All commanders and the men they command will be the focus of transformation and the man behind the machine will continue to be the nerve centre. NCW requires technology but ultimately reliance is essentially on people and organisations. An NCW environment would demand a whole new set of skills and competencies from the Information Age warriors. These warriors must have a thorough understanding of all the system capabilities present in the battlespace and the ability, initiative and innovativeness to employ the capabilities for best effects. They must have the ability to interpret and make decisions on incomplete data, or when flooded with data (information overload), ability to operate in flatter organisational hierarchies. In addition, capability to deal with lethality and accuracy of new technologies and adaptability and flexibility to cope with change are required. Although development of automated Operational Information System (OIS), Management Information System (MIS) and GIS is currently underway in the Services; managing the actual transition is a major challenge which involves both the technological as well as psychological aspects of change. There is an inescapable need for standardisation and commonality of equipment and protocols so as to achieve integration of the individual modules and systems for an integrated and resilient networked architecture.

The major test of leadership will be the ability to function in an environment of increased speed of command, requirement of self synchronisation and impact of these operational imperatives on the entire command process. Leadership will need to deal with dispersion of authority. The networked environment will present tempting opportunities to scrutinise every action of subordinates, but leaders will have to find the right balance between able leadership and micromanagement. Even when systems are linked and networked, organisational culture can often play the spoiler. Operating with flexibility and speed within the commander's intent requires a paradigm shift in tradition and culture. It needs decentralisation of authority based on trust, empowerment and confidence in the decision-makers. We need to cater for enhanced education and intellectual standards for both soldiers and officers in order to optimise the recruitment policy. Right from the nascent stages of training, aspects of network centricity must be dovetailed, and thereafter be made progressive. Exercises will need to focus more on gaining experience and familiarity with utilisation of the networked medium for interaction. The main challenge will be to train soldiers to work in small teams, without compromising the benefit of traditional unit cohesion through the joint training to facilitate interoperability. This needs to be aimed at the middle level management. Joint-Service courses need to be introduced for officers of the three Services. Inculcating inter-Services jointness should be a priority. In particular, common methodologies and joint staff procedures will have to be worked out and practised as a routine. The existing training establishments of the three Services should be developed and harnessed for network-centric training. In addition, joint training establishments need to further focus on joint operational training of personnel from the three Services. As for human resource, while we are training our personnel for system administration and operation and maintenance, it merits consideration to outsource trained manpower through lateral induction of core specialists from the civil domain. Training is a challenge in transforming the military into a network-centric force and is being addressed to some extent. However, some drastic measures are also needed to streamline procedures; prune in-Service bureaucracy and increase focus on development of human resource. For example, while outsourcing of a study to work out the IT training for the three Services was approved by the MoD some four years ago, it was shelved because of inter-Service disagreement of its necessity. It must also be understood that training not only involves new systems but also transformation in concepts and methodology of war.

[6]NCW demands the creation of a more dynamic and responsive organisation since it involves corroborating and sharing of information to ensure that all appropriate assets can be quickly brought to bear upon the combat operations. It presupposes the

---

[6] Poirier, John A; Bates, Edgar and Tempstli, Mark, 'How Much is a Pound of C4ISR Worth-An Assessment Methodology to Evolve Network Centric Measures and Metrics for Application to FORCEnet', http://www.dodccrp.org/events/8th_ICCRTS/pdf/143.pdf

abandonment of the classical hierarchical command and control system. Horizontal fusion of information calls for sharing of information by all components in the organisation in real time which requires communication networks and connecting all links simultaneously. If we are to realise this at the tri-Services level, NCW will have to be forged on suitably integrated organisations, joint concepts, joint doctrines, new technologies, joint training and joint communication architecture. Therefore, the important issues that the Services need to examine in far greater detail are joint concepts and a joint doctrine to fight future conflicts, joint/integrated organisations, joint communication architecture, induction of new technologies, network-enabled platforms, attitudinal change to accommodate the concept of NCW and an adapting military leadership to take in the changing nature of war. The NCW architectural model must consist of a network of networks with inherent redundancy. [7]The architecture must ensure that all the components are networked seamlessly, allowing the top commander of the armed forces and the political hierarchy charged with security of the nation to access all the information with regard to military, economic and diplomatic capability of the enemy. The military inputs should be provided by a functional, robust, secure and redundant network with unlimited bandwidth. The firewalls, protocols and gateways of this functional network must ensure the flow of genuine and relevant data to avoid information overload. Once the operational picture is presented simultaneously to all the strategic decision-makers, the architecture must allow seamless interaction through a network to arrive at an optimal decision. At the operational-level, the architecture must be designed to network the organisation in a manner similar to that at the strategic-level. At the tactical-level, the architecture must

---

[7]Kamradt, Henry and MacDonald, Douglas, 'The Implications of Network Centric Warfare for United States and Multinational Military Operations', Occasional Paper, Center for Naval Warfare Studies, USA, http://www.dtic.mil/cgi-bin/GetTRDoc?AD=ADA430553

ensure seamless integration (machine to machine) of all manned and unmanned platforms and weapons. The ground commander should have the information of the location of the soldier or the tank in real time on a raster map on his palmtop, which should be directly linked to the elements providing close support (artillery battery or the Air Force pilot), thus establishing a Common Operational Picture (COP). A highly reliable application network at this level is desired to establish situational awareness down to every section commander/tank commander or the pilot of the aircraft to determine the whereabouts of his troops and identify the enemy locations.

While the three Services modernise their respective networks and suitable gateways are catered for limited integration at appropriate levels, a major challenge will be that the existing communication networks do not allow the desired level of interoperability. The completion of the DCN, which is being fielded as the tri-Services strategic communication network for implementation of the C^4I^2 concepts, will lead to connectivity down to the Corps Headquarters in the Army, Maritime Operation centres of the Navy and Air Defence Direction Centre (ADDC)/airfields of the Air Force. This tri-Service architecture should be conceived well and developed after taking into account various types of operational settings derived through strategic and operational tri-Service war-gaming. A project like implementation of NCW is a multidisciplinary process. [8]It is, therefore, imperative that some of the critical issues are addressed at the inception stage. These include evolution of an enterprise architecture, integration and interoperability, communications, bandwidth requirements and latency, new technologies like Software Defined Radios (SDRs), robustness of transmission, message and signal routing, sensor exploitation, management of databases,

---

[8]'The Implementation of Network-centric Warfare', Office of Force Transformation, Department of Defense, USA, http://www.au.af.mil/au/awc/awcgate/transformation/oft_implementation_ncw.pdf

information security, information overload, integrated logistics, dangers of micromanagement, commercial influences, strategic initiatives and time for implementation.

[9]Cyber warfare is the latest and perhaps the most potent instrument of war. It provides the means to conduct attacks and weaken enemy capabilities even before the declaration of war. It is increasingly evident that China has the most comprehensive cyber warfare capabilities which have been suspected of hacking foreign government networks. As a country, we have yet to grasp the significance of cyber warfare. There is no cohesive policy and organisation for cybersecurity at the national-level. [10]The least that can be done will be for the Services to create a unified Task Force on the lines of the under-construction US Cyber Command. This should be a prelude to the setting up of a National Cyber Command, albeit progressing both concurrently is the need of the hour. We must be able to prevent cyberattacks but if they happen, we should be able to contain them and effect swift recovery. Foolproof mechanisms should be developed to check the system for malware, a capability that is non-existent in India today. Malware that is embedded in both software and hardware can prove seriously risky to national security. This is particularly relevant in a country which imports a phenomenal amount of computer parts from China. Like the US, India too, must adopt an unambiguous Offensive Cyber Warfare Policy. We must make cyber dominance an essential component of our war doctrine. In 2008, the [11]Russia-Georgia conflict became a defining event in network warfare. Open source reports that altered

---

[9]Brenner, S, *Cyber Threats: The Emerging Fault Lines of the Nation State*, Oxford University Press, USA, 2009.

[10]U.S. Army Cyber Command, http://www.arcyber.army.mil/

[11]Smith, Philip, 'How Seriously Should the Threat of Cyber Warfare be Taken?' 17 January 2014, http://www.e-ir.info/2014/01/17/how-seriously-should-the-threat-of-cyber-warfare-be-taken/

Microsoft software was fashioned into cyber weaponry and hackers collaborated on US-based Twitter, Facebook and other social networking sites to coordinate the attack on Georgian digital-based targets. For the researchers, a striking revelation was how quickly a common citizen could be transformed into a foot soldier in a cyber conflict. The cyberattacks were carried out by civilians with little or no direct involvement with the Russian government or military, and were aimed at disabling the Georgian government, banks and media outlets. We need to guard against such threats. Other than hackers/cyberattacks, our networks will face serious threats even from non-nuclear electromagnetic pulse weapons and microwave weapons much before the battle is joined. E-bombs are a real threat now. Although, electromagnetic pulse and high pressure microwave hardening by retrofitting is a very expensive process, engineering requisite resistance into a system ab initio adds little to the overall cost. Given the incapacitating potential of these weapons, we need to develop such capability indigenously.

Operational and tactical networks of the military need high mobility using a combination of terrestrial, wireless links and satellite overlays plus robust Electronic Counter-Countermeasures (ECCM). Future combat systems will be increasingly reliant on mobile broadband tactical communications with resilient wide area coverage. Internationally, high quality and rich military communication systems have already been developed and are being continuously upgraded. Lack of focused R&D in our country and the increased requirement of high capacity state-of-the-art telecommunication systems for our military communication networks have prompted the import of Commercial Off the Shelf (COTS) hardware. Adequate frequency spots and bandwidth required for optimally deploying these systems need to be made available for exclusive defence use. There is a pressing operational requirement of allocating a dedicated defence band with adequate frequency spots to cater to existing and planned defence communication systems.

The strong relationship between personality and decision-making has been proved through various studies. Personality has significant influence on general orientation towards goal attainment, selection of options, treatment of risk and reactions under stress. Multiple or contradictory intelligence inputs coupled with fog of war further complicate decision-making. A Decision Support System (DSS) with capability of advanced Multi Sensor Data Fusion (MSDF) that presents a coherent Common Operational Picture at the tri-Service level is absolutely essential. Updated and accurate knowledge of the terrain being a battle-winning factor, a superior quality GIS platform providing intimate details of terrain is needed. Since proliferation of automated networked systems will induce dependence, even the smallest downtime can result in serious consequences. Reliability and robustness with almost foolproof uptime are vital prerequisites. In-built redundancy, interconnected through self-healing networks is required. Benefits of virtualisation too, need to be harnessed to provide high system availability. Multiple surveillance resources and weapon platforms require a robust and efficient BMS that is user-friendly, not cumbersome, on-the-fly, rapidly adaptable to changing force levels and with high assurance secure communications for integrated and optimal employment of all combat resources at the point of decision. Artificial Intelligence and Robotics are finding increased applications in military use. Intelligent systems that 'learn' from past experience can prove to be very useful in military applications like Network Management System (NMS). Similarly, Robotics holds great promise for military use by way of unmanned vehicles for reconnaissance and surveillance, bomb disposal etc. Concentrated efforts are required for identifying areas in which these technologies can be exploited. Contribution by agencies like Centre for Artificial Intelligence and Robotics (CAIR) and the domestic industry is important for development. The latter will need to be forthcoming in this regard as the past record of DRDO/CAIR in these fields has been lagging, to say the least.

Technological developments in IT and communications far outpace existing procurement processes. Long gestation periods of projects risk obsolescence and delay modernisation. Further, more time required to develop security algorithms and obtain Scientific Analysis Group approval add to the problem. There is an urgent necessity to evolve a separate procurement procedure for IT and communications, ensuring faster induction without compromising transparency and cost-effectiveness. Robustness of transmission, adequate bandwidth, media redundancy and efficient routing are vital for netcentricity. Full utilisation of new technologies for development of TacC^{3}I system is adversely affected by availability of frequency spots for defence purposes in the desired frequency bands. The military should also plan on deploying new technologies like SDRs to offset growth in demand for spectrum. Effective management of a scarce resource like spectrum to assure connectivity is a major challenge in the network-centric environment. Also, for passage of information, compression technologies must be capitalised on. Robust security algorithms must be speedily developed to ensure security of both stored data as well as transmitted information. India has still very little indigenous R&D in developing state-of-the-art technologies to meet the requirements of networked systems. There is a compelling need to develop our own operating systems, GIS software, computing and networking hardware with standardised proprietary protocols and standards. The industry must go all out in development of new technologies, customised for military use and their working to operate in a synergised manner as this is the biggest implementation challenge in our efforts towards netcentricity.

In the backdrop of today's asymmetric threats and the emerging geostrategic environment, netcentricity of the military alone will not suffice. Future threats indicate national-level netcentricity as an essential requirement. Not only is information ascendancy crucial for national security, we need to establish a national information grid linked to all concerned, including the military. A national data super-highway interfaced with a national intelligence database and with need-to-know basis access rights to various defence, law enforcement and other government agencies has the capacity to revolutionise the way in which the national defence and security concerns are currently being addressed. The DCN should be linked with this national information infrastructure at the appropriate level to ensure 'Information Dominance'. The military is at the threshold of an information revolution. [12]The Services need to take a holistic view of the requirements and formulate a joint NCW doctrine which should act as a guiding document for evolution of the NCW architecture. Concentrated and focused efforts are required to integrate the system of the Services, eventually integrating them into the overall national information structure. The challenge to the defence forces in general and the IA in particular, is immense and needs to be addressed head on.

For effective implementation of the transformation process, certain imperatives at the national-level are required. First, the political hierarchy should ensure that all security related organisations must work in synergy on various aspects leading to the implementation of netcentricity for achieving the common goal of national security. Second, a joint force networked for NCW must be established and all our future accretions should be designed as 'net-ready' and interoperable with due emphasis on indigenous research and self-reliance to ensure security and redundancy in our systems. We must develop NCW-related concepts and also requisite capabilities through models and simulations. Third, realistic and quantifiable goals should

---

[12]Smith, Christopher R, 'Network Centric Warfare, Command, and the Nature of War', *CGSC*, 2009, http://cgsc.cdmhost.com/utils/getfile/collection/p4013coll3/id/2489/filename/2490.pdf

be set, developing an implementation plan to achieve these and in doing so also be able to measure the progress made. An immediate goal must be the availability of a networked joint force as a test bed that can experiment with the concepts and capabilities of NCW. Fourth, we should exploit our expertise in information technology to create an interface between the defence forces and industry with profitable pay-offs for both to foster self-reliance.

HQ IDS should be empowered to achieve true synergy and seamless interoperability in the Services. Aside from the Services-level, we should formulate a suitable joint NCW doctrine at the national-level providing an evolving strategic vision of 'full spectrum dominance'. Indigenous research and development to acquire critical technologies and innovate and absorb the same must be developed. Establishment of a robust ICT infrastructure is a critical requirement. Effective management of the transformation and increasing the technical threshold of the users is essential. Cyber warfare technology needs to be nurtured to develop the asymmetry against the adversary by denting his networks and downgrading his fighting ability. Our inherent strengths lie in the dominance that India possesses in the fields of software. We need to capitalise on this; strengthen this in-house capability to augment our technologies and enhance the effectiveness of our operations. The jurisdiction of the military, Para Military Forces (PMF), Central Armed Police Forces (CAPF), police and civil agencies is usually well defined. In practice, there is a fair amount of commonality of raw data as well as overlap of effort. Lack of networking creates avoidable gaps in information and duplication of tasks. Commercial organisations have been using some form of networking for many years. The success of these electronic solutions have always been an attraction for the defence and security communities for their own unique missions, but they have also been acutely aware that early users have had to endure frequent breakdowns in communication links, repeated software glitches and hardware problems. At critical moments of national security, there is no scope for such hiccups. The last decade has seen significant improvements in sensors, high-speed digital data transfer using worldwide space, optical and mobile telephony links and the ruggedisation of the hardware, coupled with greater affordability. Cutting edge sensor technology has ensured that the results remain unaffected by bad weather and light conditions. The important thing is that precise data and/or imagery for quick and accurate decision-making is available on call, creating battlefield transparency and situational clarity even under the most trying circumstances. History will mark the twenty-first century as the age of networking, with individual systems plugging into larger systems, thus leading to the ultimate goal of a 'System of Systems'. We must take full advantage of this and aim for complete Network Centricity for our national security.

5

C⁴I²SR

In the wake of speedy technological advancements, Command, Control, Communications, Computers, Information, Intelligence, Surveillance and Reconnaissance (C⁴I²SR) systems provide sterling opportunities to the defence and security establishment, acting as important force multiplier for commanders at all levels. Akin to any technology, undoubtedly there are problems but these are not insurmountable. The challenge lies in making C⁴I²SR systems interoperable, survivable, inexpensive, reliable and maintainable on the battlefield. [1]Given the terrific capacity building that C⁴I²SR systems offer and the effective training on these systems, our forces will have that necessary edge to emerge winners in future conflict situations, and highest level of tactical and strategic information available in order to back up military decisions and actions. Rapid developments in technology have revolutionised warfare. The key to success will lie in attaining higher levels of netcentricity, effective command and control across the force, an accelerated decision action cycle and an ability to conduct synergised operations

---

[1]Dayal, Shantanu, C⁴I²SR for Indian Army, *Bibliophile South Asia*, New Delhi, July 2012.

simultaneously within the defence and security establishment. Harnessing information technology will act as a force multiplier to enhance operational effectiveness of commanders and troops at all levels by enabling exchange, filtering and processing of ever increasing amounts of digital information presently available but not integrated. This is very relevant at all levels and particularly, at the cutting edge where opposing forces are in contact.

We are in a state of perpetual conflict. In fact, it has become a global phenomenon with many nations engaged in asymmetric wars. Within the protracted full spectrum conflict, adaptive and asymmetric threats have overshadowed conventional conflict. The changing nature of conflict has added new complexities and challenges. Conventional conflict is increasingly intertwined with irregular forces using unconventional means and tactics, while irregular forces are becoming increasingly lethal, with access to technology and equipment that previously only conventional forces used. The rapid development of high-technology weapon systems and their possession by powerful states meant that weaker states and non-state groups could no more stand up to powerful states. This led to asymmetric conflict by smaller, irregular forces employing terrorism, insurgency and guerilla warfare while exploiting technology, networks, cyberspace and media. Fighters today are tough, techno-savvy and ideologically motivated with cultural awareness impacting military operations. Launch of tactical missions has strategic implications. Information, space and cyber operations are being waged continuously. There are no rules, regulations and boundaries either. Defence forces and nations are using networks extensively and are feverishly engaged in bettering niche capabilities above adversaries. [2]Emerging conflict scenarios require a shift from

---

[2]Kuhn, James K, 'Network Centric Warfare: The End of Objective Oriented Command and Control', Naval War College, USA, 1998, http://www.dtic.mil/dtic/tr/fulltext/u2/a348447.pdf

attrition-style warfare to faster and more effective concepts of speed of command, precision and synchronisation. The new war paradigm demands integrated and networked decision support systems with space, land, surface and subsurface sensors with state-of-the-art weapons and equipment whose potential requires optimum utilisation and synergy to inflict maximum damage on the enemy.

Why C⁴I²SR?

We need C^4I^2SR to create positive asymmetrical capabilities and comprehensive competitive edge over adversaries to transform its current perception and thought. Technological advances enable precision delivery of enormous firepower of hundreds of megatonnes on a given target thousands of kilometres away. Everyone, from the man on the battlefield to the entire chain of command and control, right up to the chief political executive and the people, are on the same real time grid. The implication for the military is that while maintaining information superiority, decisions have to be taken swiftly and efficiently, and the entire consultation, decision-making process has to be radically reviewed.

Technology

Technology provides numerous options to build automated C^4I^2 systems for effective leadership in the modern battlefield. Success in combat depends greatly on fused, tailored intelligence which is communicated securely and rapidly. Speed is a crucial component. The critical elements of Sensor and Engagement Grids are hosted by a high quality information backplane. These are supported by value-adding command and control processes, many of which need to be automated to achieve speed. This in essence personifies the essential characteristics of a C^4I^2 system. Therefore, there is a need for a tremendously flexible and robust C^4I^2 architecture which functions as a process of organisations, doctrines and technologies. An important related issue is the increased role of IW, due to the exceptional growth in technology over the past few decades. IW, therefore, attempts to influence the first three activities of the OODA loop by disrupting enemy's observation and surveillance systems, corrupting enemy's orientation and misguiding his perception and finally inducing him to arrive at a wrong decision. This aspect of IW has significant implications for any C^4I^2 architecture. Information Integrity, Information Assurance and Information Superiority are the key factors while conceptualising any C^4I^2 system. The digital age promises to revolutionise the battlefield decision-making process. Here again, time is the decisive factor. There is likely to be an overload of information from a variety of sources and there will be tremendous pressure on the decision-maker to keep up with the pace at which decisions would require to be taken. In the process of this balancing act between the requirement for speed and the need for quality, there is a likelihood that the quality of decisions will be affected. It is in this context that an efficient and reliable C^4I^2 system will become indispensable. Surveillance and Reconnaissance (SR) transcends space, air, surface and subsurface means. Satellite and space surveillance apart, smaller and smarter unmanned aircrafts are transforming spying and surveillance. The nature of air power and the notion of air superiority at least in Iraq and AfPak region have been transformed in the past few years because of massive deployments of UAVs that are drastically cheaper than fighter jets. UAVs are excellent surveillance means. By operating discreetly, they can intercept radio and mobile phone communications, and gather intelligence using video, radar, thermal imaging and other sensors. The data they gather can then be sent instantly via wireless and satellite links to an operations room halfway around the world and to the handheld devices of soldiers. Northrop Grumman's Global Hawk has tremendous potential and reach. C^4I^2SR enables operators to manipulate battlefield weapons including armed UAVs from control rooms half a world away. Defence forces with such technology have great battle advantage.

Framework

Similar to the commercial sector which requires a critical mass of connectivity, computers and customers to successfully innovate e-business solutions, defence services require a similar critical mass of integrated communications and computing capability. C^4I^2SR is that critical path to transformation for the latter. The ability to conceive, experiment and implement new network-centric ways of doing business that leverage the power of Information Age concepts and technologies depends on what information can be collected, how it can be processed and the extent to which it can be distributed throughout the organisation. The ability to bring this capability to war will depend on how well it can be secured and its reliability. Security, robustness and interoperability will be key factors. The ability to protect our information, systems, programmes, people and facilities in a risk management environment directly impacts our ability to successfully prosecute the military mission. We need to develop improved methods and techniques to anticipate probable threats, ascertain our vulnerabilities and integrate practical counter-measures thus, maintaining a security-conscious workforce during concept formulation and applying effective countermeasures to the full range of systems, programmes and critical technologies. The C^4I^2SR system must be robust with sufficient connectivity and bandwidth. Access to adequate radio frequency spectrum for data transport such as satellite links, wireless networks and mobile communication systems would be essential for the system to operate effectively. Interoperability is a key parameter in all military operational and systems architectures. Retrofitting interoperability is costly, does not satisfy mission requirements and creates security problems. Ab initio joint and combined systems, achieved through engineering in interoperability attributes from the start, will provide the needed capabilities. However, this is a luxury our defence forces do not have.

The challenge is to harness the power of sensors, information processing and communication technologies to develop concepts of operation and command and control approaches that will be driven by information awareness rather than uncertainty. The ability to integrate across a number of dimensions will determine how successful we are in bringing all of the available information and assets to bear in any given situation or circumstance. These dimensions include time, echelons, functions, geography, agencies and coalitions. There is a need to assemble 'System of Systems' with co-evolved organisations, doctrines, processes and information flow that will enable this integration to occur. For example, temporal integration (such as getting the Commander's intent to all relevant subordinates at the same time) promises to result in less confusion and to reduce the fog of war while at the same time enabling a greater degree of simultaneity. The same Information Age technologies will also enable continuous Command and Control (C^2) processes, to replace the cyclical processes of the Industrial Age. Integration across echelons and functions can also reduce the fog of war and help ensure coordination of activities such as logistics, operations and intelligence. Integration across space or geography is the key ability to amass effects without the need to gather forces. Finally, integration of operations and interagency efforts is essential to achieve a unified effort, one of our most urgent challenges.

The ultimate goal of the C^4I^2SR transformation must be to transform into NCW capabilities; the development of a force that provides the commander with the capability to dominate across the spectrum of operations within the context of future security environment. Information Superiority, Decision Superiority, Dominant Manoeuvre, Precision Engagement, Focused Logistics and Full Dimensional Protection are required, with C^4I^2SR having a direct bearing on the first four. Information Superiority is fundamental to the transformation of the operational capabilities of a force and in creating a qualitative change in the information environment that will result in profound changes in the conduct of military operations. Information advantage can be effectively translated into a competitive advantage when it enables commanders and their forces to arrive at better decisions and implement them faster than an opponent can react. In a non-combat situation, this translates into the capability to make decisions at a tempo that allows the force to shape the situation or react to changes and accomplish its mission. These collective capabilities constitute Decision Superiority that results from superior information filtered through a warfighter's experience, knowledge, training and judgment. A commander's capability to achieve Decision Superiority is enhanced through the expertise of supporting staff and the efficiency of associated processes. Dominant Manoeuvre is the ability of a force to gain positional advantage with decisive speed and overwhelming operational tempo in order to achieve assigned military tasks. Precision Engagement is effects-based engagement that is relevant to all types of operations. Its success depends on in-depth analysis to identify and locate critical nodes and targets. The pivotal characteristic of Precision Engagement is the linking of sensors, delivery systems and effects. NCW concepts and capabilities effectively network sensors, command and control and shooters to engage with precision across the depth and breadth of the battlespace.

When looking for technological solutions for C⁴I²SR, we should examine the requirements of collection, exploitation, storage, retrieval, distribution, collaborative environments, presentation, information operations and information assurance and the technologies that help extract knowledge and understanding from data and information. Our focus should address Connectivity, Technical Interoperability, Semantic Interoperability, Integrated Processes, Integrated Protection and Networked Battlespace Enablers. We should understand that the potential value of the defence force is the sum of the potential value of its entities, which in turn is heavily dependent on the nature C^4I^2SR that connects them.

Organisation

C^4I^2SR systems should enable warfighters at the lowest levels to know as much as the most senior commanders do about the combat situation throughout an entire battlefield of operation. The result of this information networking should be a decentralisation of command authority with individual warfighters empowered as never before. It would offer unprecedented opportunities for initiative and independent operations by individuals and small units. C^4I^2SR should provide the networked organisation wherein the information gathering process will be more equally distributed and more information will be available rapidly to all levels of command. Commands will share rather than control information, resulting in faster decision-making at all levels of command.

Communications

C^4I^2SR will essentially enhance the quality of decision and speed of command as well as quality of information along with shared situational awareness, ensuring greater collaboration and flexibility. The system must include robust and secure communications as fast paced, precision and decisive operations will need to be executed, particularly at the operational and tactical-levels. C^4I^2SR communications should be balanced so that over-dependence on technology and inhibitions about combat operations are avoided. The design should suit information sharing needs of all entities. As mentioned earlier, a 'System of Systems' approach should be followed, which will simplify the complicated work of data gathering, processing, fusion and management. By integrating several subsystems, one logical network can be created. The communications, thus, achieved would be seamless and from anyone to anywhere. Information exchange should be designed in a manner that suits the need of all entities; identifying information heads to data linkages. C^{4}I^{2}SR must be available throughout the spectrum of conflict, ensuring smooth communications and operations in all geographical and climatic conditions. Manoeuvre necessitates that C^{4}I^{2}SR communication systems are largely on wireless. Therefore, large networks would need to be synchronised and alternatives explored to overcome challenges. Disproportionate increase in interoperability and interdependency, increased emphasis on end-to-end functionality rather than individual component capability, loosely coupled systems instead of specific task-oriented system are some of the challenges in future architecture. The communications architecture should also include extensive use of satellite and micro UAV-based communications. Significantly, while communications is the key to netcentricity, bandwidth is the driving factor. Bandwidth capacity of wireless networks will have a fundamental bearing on net-centric operations. Communications infrastructure is the key to NCW. [3]As per Metcalfe's Law governing network-centric computing, the power of a network is the square of the number of nodes in the network. This power can only be harnessed if matching mobile communication with requisite bandwidth is provided to integrate the nodes to the network in time and space. Lack of communications/ interim arrangements should not hijack networking. Bandwidth is important. The US had forecasted 1 GB per sec requirement of bandwidth for a single combat team by 2012. We need to examine use of commercial satellites for military communications with adequate security superimposed, as is being done by developing countries. Lack of bandwidth on wireless media for mobile operations is one of the most complex technological challenges to NCW. A potential aid is the use of computer software to compress signals to ensure least consumption of space in the frequency spectrum. Higher capacity Software Defined Radios (SDRs) that have the ability to

---

[3]Metcalfe's Law, http://www.princeton.edu/~achaney/tmve/wiki100k/docs/Metcalfe_s_law.html

transmit large volumes of information in lesser bandwidth than traditional radios need to be inducted. Services must continue to leverage commercial technology to determine alternate solutions to solve bandwidth problem. In the face of transmission impairments such as solar flares, bad weather and hostile jamming, networks must continue to function. If a signal cannot penetrate a rain shower or is blotted out by enemy barrage jammer then the link is broken. This would render the network model non-functional. Platforms must be able to specifically address and access other platforms or systems in the networked environment, coping with a fluid network topology as platforms enter and leave an area of operations frequently. The answer lies in establishing 'Plug and Play Connectivity'. The future demands deployment of large number of sensors over vast frontages that need complex application data fusion. It would also demand management of various communication links, using diverse protocols across the frequency spectrum. Communications concentration, data management and fusion remain critical. The big challenge is to design an application with advanced artificial intelligence capability to detect and track legitimate targets. Collection, collation, interpretation and dissemination of information in near real time can achieve information superiority. Data capture and their processing, including data fusion, hold the answer to this force multiplier.

Security

The potential for compromise is inherent in network warfare. The more data you share, the greater the chance of compromise. There may be some virtue in keeping some elements on the net, and retaining tight hierarchical control of certain critical need-to-know elements. The level of active and passive security measures that are instituted will dictate success levels in future battles. We need to be alive to 'Net Forces' in our neighbourhood and microwave weapons being developed. Capability to wage cyberwar by denying information to the enemy, while launching active attacks on enemy networks and databases would give the power of information dominance. These are evolving capabilities and take years of toil to achieve credible capability. Our neighbourhood is making giant strides in these fronts and we must work on the antidote. Significantly, the US cyber warfare strategy is 'offence being the best form of defence'. We should design our own operating systems, create redundancy of networks and significantly enhance our chip manufacturing capabilities. Passive measures would include deception, disinformation, dispersion, National Broadcasting Company (NBC), electromagnetic pulse and high pressure microwave hardening. Such hardening is undoubtedly very expensive if one were to go for retro-fitment but if introduced ab initio, costs would go up marginally. Not everyone understands the difference between cybersecurity and Information Assurance. Cybersecurity is only one part of Information Assurance. We need to expand on the base of cybersecurity to ensure full-fledged Information Assurance. Implementation of NCW will exponentially increase number of information sources and volume of information flow. Without measures to mediate this volume, information overload will occur much more than in the past. Increased information complexity can cause loss of situational awareness or unmanageable increase in mental workload. Of particular concern will be how to clarify, filter and synthesise enormous amounts of information generated by NCW sensors to take critical decisions under time pressure and uncertainty. Automation can aid, in that it can quickly sort, filter and optimise a set of multi-objective functions.

Tri-Service Scene

C⁴I²SR in the Indian military is developing in stand-alone mode of individual Services and the capabilities are yet to be integrated. The capabilities are more in Navy and Air Force than the Army. Basic C⁴I²SR infrastructure in Navy and Air Force is in place and undergoing upgrades. However, integration of disparate systems of the three Services is yet to be achieved. Procurements and developments continue on single Service basis. In the recent past, even imports like UAVs were not central to the Services. Even today, exchange of UAV pictures from one Service to another is not possible because of lack of an understanding among Services or a handshake, that should have happened years back, had we ensured a top-down approach. In absence of this, UAV pictures of individual Services continue to come up on separate screens in the Interim National Command Post (INCP) which indicates the poor progress made by us over the years towards attaining NCW capabilities. Similar is the case in operational Command Headquarters of the Army where the Air Force picture does not come directly into the Operations Room of the Army. This is just one example of lack of networking that deters optimisation of information and overall resources in the battle zone in real time or near real time.

Mention 'defence' and you visualise the wholesome construct of the Army, Navy and Air Force but any talk of 'defence communication' simply boils down to DCN, which is only a very small part of how defence communications need to be strategised, synergised and developed. In the present dispensation, defence communications are a neglected sector notwithstanding the hype of information warriors in the making. Not only are communications of the three Services mismatched, little thought is being given to the required synthesis of 'information' and 'communications'. The concept of ICT is yet to take off. As mentioned in the previous chapter, Project DCN still few years away, will simply link the three Services down to Corps Headquarters (Corps HQ) and equivalent levels plus a few static entities. Most significantly, the project does not include development of requisite software. The vital handshake, therefore, is missing without which DCN actually boils down to a highway, sans traffic. This denotes a lopsided approach and indicates the low priority accorded to a strategic communications project like the DCN.

In the Second Lebanon War, Special Forces of the three Services of Israeli Defence Forces discovered after being inducted into Lebanon that they could not communicate with each other. In the Indian context, such mismatch will be in entirety for our own Indian Defence Forces; inter-Service communications will remain at the level of Services HQ unless corrective actions are taken. Without such actions, information warrior of one Service will talk to his counterpart in another Service by going vertically up to his Service HQ, horizontally to the next Service HQ and then down to his counterpart. This will happen only if the inter-Service formats are standardised, which is still a distant future. Jointness in the three Services today is largely a misnomer. The battlefield of tomorrow will be a non-linear, multi-dimensional battlespace characterised by nuclear ambiguity, increased lethality, a very high degree of mobility coupled with simultaneity of engagement and increased tempo of operations with compressed time and space coupled with a high degree of transparency. In our context, the nuclear factor will further limit the depth and duration of conflict which would be short and intense. This would call for a swift and concerted response by the three Services, coupled with quick decision-making, the framework of which would already have to be in place, with a joint command and control structure to direct the operations.

That a single Service cannot cope with future scenarios is an acknowledged fact. This is relevant not only to conventional war but spans the complete conflict spectrum plus disaster management, Out Of Area Contingencies (OOAC) and the like. For the Army itself, future military operations will be combined and joint comprising all arms and inter-Service elements. These operations will require units and subunits of other arms to operate subordinated or in cooperation with each other. Mission-specific link up with sister Services will optimise combat potential even at tactical-levels. Jointness is essential to military success. Success in war has been contingent on the common sense idea of jointness as seamless integration. Jointness implies achieving higher joint combat effectiveness through synergy from blending particular Service strengths on mission basis. In non-traditional terms of military strategy and doctrine, some term it as a response to the evolving nature of warfare. Given the fact that in the future all the Services will have to operate jointly, even in smaller contingencies, jointness enhances the increasing synergy of modern military forces, i.e. complementary operations built around a key force (instead of a key Service). In other words, no single weapon or force reaches its full potential unless employed with complementary capabilities of the other Services. In this milieu, smooth and real time communication will be the battle winning factor.

Our military has yet to fully realise the essential requirement of viewing information from the strategic viewpoint and recognise it as a mission-critical resource. This requires a synthesis of communications and information. In its concerted efforts to modernise, the military must align with this truth and stop treating information just as another resource. Communications cannot be planned in isolation anymore. It must be part and parcel of the C^4I^2 package in concert with our pursuit of NCW capabilities. Significantly, the planned reorganisation of Pakistan Army's General Headquarters (GHQ) envisages merger of the Communication and Information Systems branches. The PLA is getting informised at a fast pace and the Pakistan Army is following suit with focused investments in relevant sectors including communications and information.

Indian Army's capacity building in C^4I^2SR comes up for discussion time and again while focus on similar capability of the Indian military remains peripheral. The reason for the latter is essentially that HQ IDS has come up as a separate HQ instead of being integrated with the MoD and sans any worthwhile clout in absence of a CDS having been appointed. The TacC^{3}I, which essentially is the mainstay of the Army's C^4I^2SR, has been inordinately delayed for years without anybody taken to task. The Defence Procurement Policy 2013 does not match fast-paced technological changes, particularly for information and communication systems. Then, there is a lack of overall focus and understanding of technology. The urgency of time factor appears lost, indicating that the importance of information as a strategic asset is understood in ambiguous manner.

Army's TacC³I

The Army's TacC³I comprises the ACCCS, ADC&RS, BSS and BMS, all of which have been/are being developed directly under the DGIS. These will be integrated through the CIDSS, also being developed by the DGIS. The TacC³I will also integrate the IA's Electronic Warfare System (EWS) and Electronic Intelligence System (ELINT) operation under Military Operations and Military Intelligence, respectively. The TacC³I is to provide state-of-the-art C^4I^2 connectivity within the IA at Corps HQ and below levels. Upward connectivity from Corps HQ to Army HQ-level is to be provided through the Army Strategic Operational Information Dissemination System (ASTROIDS), also being developed under the aegis of the DGIS. The current status of these various systems is as follows:

ACCCS

This is the first Operational Information System (OIS) that has been fielded into the IA that provides complete automation of artillery tasks from the command post. This project was sanctioned in March 2002 for developing a networked solution for automated tactical and technical fire control by artillery at all echelons. The orders were placed with Bharat Electronics Ltd (BEL) in July 2002, after being approved as 'Buy and Make' project. In acquiring the system, the COTS approach was followed; tactical computers were procured from Elbit, Israel for the test bed with provision of Transfer of Technology (ToT). Other hardware was obtained by BEL commercially from indigenous sources or manufactured by them. Phase 3 fielding has been completed and continuous feedback is being obtained for improvement. However, Phase 4 of ACCS has not been fielded yet for lack of security solution and upgrades required for hardware have not been finalised yet.

CIDSS

Being developed by BEL, this project was sanctioned in May 1999. Development of a Test Bed Corps comprising one Corps HQ, one Divisional HQ, three Brigade HQs and nine Infantry Battalions was to be completed in Stage 1 of Phase 1. However, only a truncated test bed could be undertaken, that too in phases, since the TCS has not been fielded. There were also other constraints like complications in porting of security solutions besides lack of requisite communications in the Tactical Battle Area (TBA). Stage 1 of Phase 1 was completed in September 2007 as per revised PDC (eight years after the project was sanctioned). Further development of the system for Stage 2 of Phase 1 (equipping and integrating remaining Infantry formations and units of the Test Bed Corps and Phase 2 of project, extending CIDSS to other arms and services of Test Bed Corps, integration of CIDSS with other components of TacC^{3}I system and equipping one designated Strike Corps) was running way behind schedule. Eventually, a contract with BEL was concluded in March 2011 for ₹1,035 crore for equipping CIDSS along with a second contract of ₹2,635 crore for the BSS. This was possible mainly because of the dynamism of Major General (now Lieutenant General) Rajesh Pant, who was then additional director general, Information Systems. However, these contracts have not been taken to their logical conclusion in the required time frame. The application is still under development. Though both CIDSS Phase 1 and BSS Phase 1 were developed by BEL, differences in perception, debate over hardware configuration and insistence on having common GIS application in both CIDSS and BSS has delayed development. Linking integration with that of a 'common GIS', without developing standard protocols and exchange formats has further complicated the issue. Going by past experience, the application is unlikely to be ready by June 2014 as claimed by BEL. Fielding of the CIDSS pan Army will most likely now take another seven to eight years. Being the hub of the TacC³I, this will delay any measure of net-centric capability in the Army. As per the already delayed PDC in 2008, complete fielding of CIDSS pan Army was to be completed by 2017.

BSS

This project was conceived to develop an automated BSS with dedicated intra-communication, which involves integration of surveillance sensors at division and corps-level on a customised GIS platform with multi-sensor data fusion undertaken at the Surveillance Centre for providing inputs to the CIDSS. The requirements at the Brigade-level were included later in 2008. Phase 1 involved provision of the concept by developing a test bed system, which has been completed and accorded operational validation. The system was developed on turnkey basis by BEL in collaboration with CAIR. Phase 2 of Project Sanjay involves equipping all corps of the Army after successful completion of 'proving phase'. The responsibility for development of the system in Phase 2 is also with BEL and despite a contract having been concluded in March 2011 for ₹1,035 crore (as mentioned above), development has been unduly delayed since it is linked with the application of the under development CIDSS. There are delays in procurement of hardware (data radios, tactical computers) plus the Army's reluctance to sign 'End User Certificate' and delay in selection of a 'Common GIS'. Along with the contract for equipping the CIDSS, a second contract of ₹2,635 crore for the BSS was also signed but these contracts have not been taken to their logical conclusion in the required time frame. As per the already delayed PDC in 2008, the complete fielding of BSS pan Army too, was to be completed by 2017. In the development of the BSS that is to integrate IA's total surveillance resources, BEL has been facing the same problems as it had in developing the ACCC—bulk imported hardware and technology but limited indigenous capacity in applications, design and software customisation.

ADC&RS

The contract for this project was signed with BEL in March 2008, with the entire project scheduled to spread over the Army's 11th and 12th plans. The project was based on the Army's philosophy for Air Defence Control & Reporting System. As per the initial plan, the fielding of the test bed was to be done in December 2009. However, this has yet to materialise. Moreover, the initial test bed was planned as independent to be integrated with Indian Air Force's (IAF's) Air Defence System (also under development) at a later stage—a folly recognised and rectified only in 2008. An integrated test bed was first planned in 2009, which was then delayed to October 2012 but has still not materialised. Presently, beta testing is being done and hopefully, the system will go for user trials by end 2013/early 2014. As per the original plan, fielding of ADC&RS pan Army was to be completed by 2014, which may now happen only by 2019-2020.

BMS

Project BMS was envisaged to enable a faster decision process by commanders at all echelons, encourage better decision due to reliable operational information provided in real time and have the ability to quickly close the sensor to shooter loop by integrating all surveillance means to facilitate engagement. This would be done through an automated decision support and command and control system, exploiting technology for mission accomplishment in the TBA by rapid acquisition, processing and transfer of information, enhanced situational awareness, capability to react to information, sharpen ability to synchronise and direct fire, plus establish and maintain overwhelming operational tempo. The system customised to the specific Army requirement, needs to be first integrated and tested in a controlled environment for which a test bed laboratory will be needed. After testing in laboratory conditions, validation trials of the system will be carried out in the field. After successful validation of the system in the field, the process for equipping will begin. The Army was late in conceiving this system; planning for NCW capabilities below Brigade HQ-level was not thought of along with other OIS. The BMS will comprise a tactical handheld computer with individual warfighter and tactical computers at Battle Group HQ and combat vehicles, enabling generation of common operational picture by integrating inputs from all relevant sources through integrated use of a high data rate GIS and GPS. Phase I of Project BMS comprising test bed laboratory and field trials at test bed location of one Combat Group and three Infantry Battalion Groups by 2012 was delayed by almost three years due to indecision within the Army concerning delimitation between the BMS and the [4]Futuristic Infantry Soldier As a System (F-INSAS) under development by the Infantry and concurrent fallout in reordering of the feasibility study. The Infantry insisted on handling Phase 3 of F-INSAS (computer and radio subsystems plus software integration) by themselves while Directorate General of Information Systems (DGIS) was already developing the BMS, including for Infantry. Ironically, the F-INSAS is still going ahead with parallel development, though the BMS design caters for lightweight, ergonomics and long-range communication over portable SATCOM (team/troop leader-level) and sensor integration is integral to the project. Eventually, end 2011, the Defence Acquisition Council (DAC) approved the BMS as a 'Make India' project, following which an Integrated Project Management

---

[4] Janu, Raveen, 'F-INSAS: India's Future Soldier', Centre for Land Warfare Studies, New Delhi, India, 21 September 2012, http://www.claws.in/index.php?action=master&task=1219&u_id=187

(IPM) study was completed. The Expression of Interest (EoI) was issued by the Department of Defence Production in November 2013 to BEL, Electronics Corporation of India, Computer Maintenance Corporation, ITI, domestic private-sector major Tata Power SED, Rolta India, Wipro, Larsen & Toubro, HCL Infosystems, Punj Lloyd, Bharat Forge, Tata Consultancy and Tech Mahindra. It is now for the industry to take up the challenge. Thereafter, it shall be possible to shortlist two Developing Agencies (DA), by end 2014. Subsequently, the design phase could commence by July 2015; limited prototype tested in laboratory by December 2016 and finally, prototypes developed and fielded for user evaluation by December 2017 (instead of earlier schedule of 2012). The cascading effect has already delayed completion of Phase 2 (Equipping) from initial plan of 2018 to 2022 and Phase 3 (Change Management and Upgradation of System) from 2023 to 2027 as per current status. This schedule is possible only if there are no hurdles including DRDO and BEL's efforts to prevent BMS from going to private industry. With experience of capabilities of DRDO and BEL in ACCS, BSS and ADC&RS, MoD should actually select both DAs for BMS prototypes from the private industry. This would be in line with DPP 2013 opening to the private sector. Apparently, DRDO and BEL are biting more than what they can chew. The BMS is a finance intensive project and exact financial implication can only be holistically worked out at the end of Phase I. The approximate cost of Phase I of the system was earlier estimated to be around ₹350 crore, which may now double up. Similarly, the overall cost of the project of fielding the BMS in the Army may jump from initial estimates of ₹23,000 crore to ₹70-80,000 crore approximately, or even more.

EWS & ELINT

The Electronic Warfare System (EWS) and Electronic Intelligence System (ELINT) are under the purview of the Directorate General of

Military Operations and Directorate General of Military Intelligence respectively. The concept of integration of EWS and ELINT through the CIDSS was approved by the Army some years ago, and implementation is likely to begin when Phase 2 CIDSS is developed. An integration exercise with CIDSS Phase 1 was undertaken under field conditions but EWS and ELINT were not part of it and only limited message transfers were undertaken. However, as per CAIR, a meaningful integrated framework is minimum two years away. Integration requires adopting common standards, structures, exchange formats and protocols. While the DRDO has developed Indian Military Message Transfer Format (IMMTF) by outsourcing it to Tech Mahindra, the same is no more than a glorified postman. We are yet to develop/create a common symbology for digital environment (conventions, attributes, linkages), data model structures for GIS ready maps or identify/develop a file exchange format that can cater to both spatial and attribute information. There are two components of GIS data: Spatial information (coordinate and projection information for spatial features) and attribute data. Spatial information describes the physical location of objects and the metric relationships between objects. Attribute data is information appended in tabular format to spatial features. Attribute data also provides characteristics of the spatial data.

ASTROIDS

This is a GIS-based application that was taken up for development by the DGMO. It is a Wide Area Network (WAN) connecting Army HQs to operations rooms of Command and Corps HQs. The network in Delhi connects important directorates at Army HQs, Defence Imagery Processing and Analysis Centre (DIPAC) and Centre for Automated Military Survey (CAMS) on Optical Fibre Cable (OFC) based LAN. Phase 1 was sanctioned in 1995 with Institute of System Studies and Analysis (ISSA), with the DRDO as the development agency at a cost of ₹10.75 crore and was implemented by DGMO but due to certain shortcomings in the software the project could not achieve the desired aim. The test bed approach had not been adopted, there was no security overlay, it was not user-friendly and the documentation was incomplete. Hence, it had to be shut down in March 2003. Phase 2 was then sanctioned in November 2003 for ISSA to upgrade the software. The project was coordinated till November 2006 and thereafter, transferred to DGIS. The PDC for the new software was November 2006, which was later extended to December 2008. The security algorithm was to be developed by the DRDO. After evaluation of software application in the laboratory it was to be fielded in test bed environment along with the security algorithm, and hardware of the nodes was to be upgraded after successful completion of field trials. Thereafter, the network was to be extended to all Commands and Corps by December 2012. However, none of this has happened because ISSA has not been able to develop the required software. This, despite the liaison with the ASTROID Induction Cell (AIC) under the DGIS. Hence, the project has been foreclosed by the Army. A Request for Information (RFI) for a fresh project is now under preparation. Hopefully, the hierarchy will understand that the solution lies with the private industry alone.

Tactical Communications System (TCS)

Although the Army has a complete corps nominated as a test bed, none of the OIS could be tested in the manner that was envisaged—at a full corps-level. This was because of the lack of a TCS in the TBA. [5]The TCS that had been approved thrice by the defence minister in the past should have been fielded into the Army in 2000. The fact

---

[5]Kanwal, Gurmeet, 'The Imperative of Modernising Military Communication Systems', Institute for Defence Studies and Analyses, New Delhi, 16 February 2010, http://www. idsa.in/idsacomments/TheImperativeofModernisingMilitaryCommunicationsSystems_gkanwal_160210

that this happened should become a case study in bureaucratic red-tapism, lackadaisical approach, lack of understanding of technology and not recognising the value of communicating information. Ironically, no heads rolled. The adverse effects of truncated test beds for information systems are apparent as these can only lead to avoidable problems coming up at the fielding/equipping stage that could have been corrected in the test bed itself. In turn, this would also lead to avoidable additional costs that may accrue in requiring upgrades immediately after fielding these systems.

[6]Modernisation of the Army has been severely affected by the lack of the TCS. Since 2002, MoD has been vacillating on categorisation of the project under Make (Hi-Tech Systems) and Make (Strategic, Complex and Security Sensitive Systems), since private sector participation is allowed in the former category and not latter. Also, to classify it as former category was attributed to the secrecy of the 'frequency hopping algorithm' contained in a tiny microchip. However, now BEL and a consortium of L&T, Tata Power SED and HCL Infosystems Ltd have been reportedly selected by the government. This is the first project under the 'Buy Indian, Make Indian' clause introduced in the Defence Procurement Procedure in 2011. The government will pay 80 per cent of the development cost while 20 per cent will be funded by the industry. For TCS, both the selected parties will make the prototype system and the best bidder will then execute the whole project. The TCS is vital for operational preparedness and force multiplication endeavour. Decisive victory in future conflicts will be difficult to achieve without robust and survivable communications, both in the strategic and tactical domain.

---

[6] Janu, Raveen, 'India's Network Centric Warfare Programme-Tactical Communication System Part II', Centre for Land Warfare Studies, New Delhi, 23 April 2013, file:///C:/Users/tara/Documents/TCS%20Part%20II.htm

MIS

Management Information System Organisation (MISO) of the Army is developing the Management Information System (MIS). It is an Army intranet-based application presently under user trials. However, its security clearance is pending, which is likely to remain a problem till appropriate security solution is found.

F-INSAS

The BMS and F-INSAS programmes are progressing concurrently; BMS under Information Systems and F-INSAS under the Infantry. BMS was conceived at battalion/regiment-level pan Army (including for the Infantry) and comprises communication, non-communication hardware and software. The lowest level to which the system will be connected is the individual soldier/weapon platform and the highest level will be with Battalion/Regiment Commander. The system will be further integrated with the TacC^{3}I system through the CIDSS. The DGIS is charged with facilitating transformation of the IA into a dynamic network-centric force, achieving information superiority through effective management of information technology. Quite logically, Phase 3 of F-INSAS (Computer subsystem, radio subsystem, software and software integration) should be a part of the BMS, especially when OIS for the artillery and air defence are being developed under the DGIS. However, the Infantry plans to develop Phase 3 (Computer subsystem, radio subsystem, software and software integration) of F-INSAS independently and not be part of BMS. A separate project of software and communication integration by Infantry will be retrograde, delay overall netcentricity pan Army and incur additional avoidable costs, especially when BMS already caters for requirements of the Infantry. It amounts to not only 're-inventing the wheel' but will require yet another project to integrate the F-INSAS with the BMS. Foreign armies have faced similar situations and we need to learn from their mistakes. Ironically, this situation was created by a former Director General of Infantry and the concerned Deputy Chief of Army Staff, who belonged to the same Infantry Regiment, took such a decision without realising the implications. Now it has become a prestige issue for the Infantry. The project has not even been categorised yet but they plan to seek approval for going along the DRDO-PSU route for development, separate from the BMS. Hopefully, the Army and MoD (Defence Finance) will scrap such infructuous re-inventing of the wheel when the BMS is already catering for Infantry requirements.

Surveillance

[7]The Indian military is expected to induct radars worth over US $8.5 bn in the next decade as a host of defence public sector units (DPSUs) as well as local and international private sector contribute to its growing demand. Various indigenous developmental projects for radars and associated equipment as well as international acquisitions are taking place. The indigenous projects include development of Active Electronically Scanned Array (AESA) radar to be fitted on the proposed Light Combat Aircraft-MK II as well as a unique 'Through Wall Imaging Radar'. Both these radars are being developed by Electronics and Radar Development Establishment (LRDE), a Bangalore-based DRDO lab. Besides, India has initiated integration of the indigenously-built Airborne Early Warning and Control (AEW&C) system with the Brazilian Embraer EMB-145 I aircraft which India is acquiring. The EMB-145I aircraft has been modified to carry the Indian-made Active Array Antenna Unit (AAAU) mounted atop the plane's fuselage. In addition, a new generation of multi-function radars which can be integrated with any weapon system to provide surveillance, early warning, interception guidance and raid assessment are also being developed. These include a medium power radar (Arudra), a low-level transportable 150 kilometre radar and a

---

[7]Arnet, Eric, Military Technology: The Case of India, SIPRI, http://www.sipri.org/yearbook/1994/10

synthetic aperture radar. These radars will be broad in size so that they can be integrated into any weapon system.

DRDO has also developed/is developing 3-D radar systems. The Central Acquisition Radar (CAR) is for use with Akash surface-to-air missiles and is capable of tracking 150 targets; two variants have already been developed. The 'Rohini' radar is the IAF variant and the 'Revathi' is for the Indian Navy. A third variant, known as the 3-D Tactical Control Radar for the Indian Army is also being produced. According to the media, US defence and aerospace major Raytheon is also talking to the IAF regarding airborne surveillance and reconnaissance radars that would be used along India's borders. Raytheon has received two RFIs from the IAF but India has not decided whether to go for an AESA system or a Mechanically Scanned Arrangement. Meanwhile, the Indian Navy has issued an RFI for 3-D radars to enhance the surveillance capability of warships. The 3-D radars will be deployed on ships which are more than 3,000 tonnes to provide 360 degree surveillance to detect aircraft, helicopters and incoming anti-ship missiles.

On 20 April 2013, RISAT-2, a radar imaging satellite, was put into orbit by Indian Space Research Organisation (ISRO) through the Polar Satellite Launch Vehicle (PSLV-C12). This is a surveillance satellite which will keep a watch on the country's borders. RISAT-2 was developed by ISRO in collaboration with the state-run Israel Aerospace Industries. The satellite can see through clouds, darkness, fog and even a few inches deep into the soil. The advanced capabilities are on account of its Synthetic Aperture Radar (SAR) that uses microwave radiation to 'see'. Only a few other space leaders such as Canada, the European Union and the US use this technology. RISAT-2 will enhance our capability for earth observation and surveillance across the border. Its coarse resolution is ten metres and its finest is one metre. But it is not good enough to see human movements or track terrorists.

Notwithstanding the above, according to Stockholm International Peace Research Institute, our Prithvi and Agni missile programmes too, should have had integrated surveillance capabilities. The Prithvi battlefield support missile's role can be expected to be similar to that of the US Army Tactical Missile System (ATACMS), but it is less flexible, being limited by the decision to use liquid fuel and the IA's limited battlefield surveillance capabilities at the missile's full range. Strictly speaking, the Prithvi system should include an integrated surveillance and mission planning support capability and is incomplete without one. While its role may be similar to that of the ATACMS, the Prithvi missile is more closely comparable to the Soviet Scud-B or the German V-2.

No new radars and UAVs have been inducted by the IA. The move to identify and induct Micro Air Vehicles (MAVs), that have already emerged as veritable force multipliers in other armies, has not progressed much predominantly since the Infantry has been focused more on the F-INSAS. Meanwhile, the DRDO is designing a range of MAVs (Black Kite, Golden Hawk and Pushpak have already been developed) as also there is beginning to be a greater availability of other indigenous products in the market like the Netra by Idea Forge, a spider-like MAV suited for all types of operations, including counterterrorism and counter-insurgency or the MAV with an infrared sensor developed by Aurora Integrated Systems.

Implementation of C⁴I²SR

C⁴I²SR systems should optimise state-of-the-art science and technology to integrate widely dispersed decision-makers, situating and targeting sensors, weapon platforms and field forces into a highly adaptive, comprehensive 'System of Systems' to achieve unprecedented mission effectiveness. For exploiting science and technology and to apply a technology-oriented framework, it should have a combination of Surveillance, Information, Command and Engagement grids. The Surveillance Grid rapidly generates battlespace awareness and self synchronisation. The Information Grid is a high performance network that provides the backplane for computing and communications. The basic level requirement for NCW is a high performance National Information Grid (which should be part of the Global Information Grid) that provides a backbone for computing and communications across the spectrum. But this remains only a potential capability until we develop the appropriate software, capture required data, have the organisational set-up, human resource and doctrines to exploit this capability. The Command Grid should also include knowledge-based artificial intelligence and software applications that act as command advisers and are able to recommend courses of actions. The Engagement Grid exploits the awareness and translates it into increased combat potential. The Sensor Grid observes, the Information Grid orients, the Command Grid decides and the Engagement Grid acts. To achieve mission objectives, the grids must interact and exchange information. It, therefore, emerges that NCW is a multidisciplinary process that can only be implemented by harnessing the available information and communication technologies. Implementation of such concept naturally lends itself to challenges that we need to overcome.

C^4I^2SR is the threshold of NCW capability, implementation of which must overcome challenges encompassing the technical, cultural, organisational and administrative domains. Likely impediments would include: One, lack of secure, robust connectivity and interoperability; two, intolerance to innovations (resistance to change); three, resistance to transparency of own operations to higher commanders; four, lack of understanding of key aspects of human and organisational behaviour concurrent to required changes; five, lack of technology understanding/investment. To overcome such impediments, implementation has to be a multidisciplinary effort while executing it in incremental fashion. This will also enable a faster user feedback and subsequent design refinements.

It is essential that a coherent framework for a 'Joint Services Enterprise Information Architecture' is defined which would support seamless flow of information across the battlefield from the lowest echelons to the strategic-level, but on a need-to-know basis. This should encompass the System Architecture, the Technical Architecture along with the Operational Architecture. The System Architecture implies what is wired to what. The Technical Architecture indicates how interfaces are defined and the Operational Architecture shows how data flows. The technological challenge is how to integrate the desired information and to speed up and optimise information sharing among systems that were not originally designed to talk to one another. The organisational challenge is in determining how to adopt new practices that cut across organisational boundaries and harness collective expertise. A major technical challenge is that while Services are modernising their respective networks with suitable gateways for limited integration at appropriate levels, existing communication networks do not allow desired level of interoperability. The DCN being fielded as tri-Services strategic communication network for implementation of the C^4I^2 concepts does not include common software. C^4I^2SR assists transition from platform-centric warfare to NCW, implying major changes in terms of manoeuvre, mass, firepower and logistics. The latter's battlespace is flatter, synchronised, compressed, interactive, with fluid boundaries, open access to information and coordinated data collection. The network infrastructure enables shared situational awareness and decentralised planning and execution with potentially greater ability for rapid adaptation and improved speed of command. Essentially, there is a paradigm shift in the battlefield. The erstwhile Decision Support System is transforming the Decision Assist System, with increasing dominance of intelligence and computers to tackle the complexity of the system. In networked environment, decision-makers, sensors and shooters work collaboratively and consistently in response to the dynamics of the battlespace to achieve the commander's mission. It is not narrowly about technology, but broadly about an emerging military response to the Information Age and technology which if misapplied within an organisation only guarantees failure.

Transformation requires alterations in our concept of operations, doctrine, organisation and force structure, and above all in the psyche of the fighting man, including the leadership. Associated changes in logistics, education and training will also be required. These changes will have to be concurrent and on existing structures so as to bring about a graduated increment in NCW capabilities within the constraints of development and implementation time. A number of steps need to be taken. These include a phased shift in existing technology at Services-level and horizontal fusion in the armed forces at laid down hierarchical structure. Human resource development must be pursued vigorously. Concerted efforts have to be made at all levels to conceptualise, define, induct, implement and operate the new systems in a defined period of time. This is vital so as to be able to contain and limit obsolescence, cost escalation and depletion of carefully created expertise and continuity. Industry support can only be expected or sustained if projects are executed professionally in a time-bound manner. Therefore, the procurement cycle needs to be shortened.

Asymmetric wars need a national response. C^4I^2SR architecture, therefore, should encompass the entire defence, security, intelligence and related government and non-government agencies including decision-makers. For effective implementation, a number of imperatives will have to be taken up at the national-level. All security related organisations must work in synergy on various aspects leading to the implementation of netcentricity if we are to achieve the common goal of national security. These changes have to be driven from the top national leadership. The architectural model for C^4I^2SR must consist of a 'networks of networks' with inherent redundancy.

It must ensure seamless networking of all components, allowing the top leadership to access all information with regard to military, economic and diplomatic capabilities of the enemy. Military inputs should be provided by a functional, robust, secure and redundant network with unlimited bandwidth. The firewalls, protocols and gateways of this functional network must ensure the flow of only genuine and relevant data, to avoid information overload. Once the operational picture is painted simultaneously to strategic decision-makers, the architecture must allow seamless interaction through a decision network to arrive at an optimal decision. At the operational-level, the architecture must be designed to network the organisation in a manner similar to that at the strategic-level. At the tactical-level, the architecture must ensure seamless machine to machine integration of all manned and unmanned platforms and weapons. A highly reliable application network at this level is desired, to establish situational awareness down to the lowest commander or the pilot of the aircraft to determine whereabouts of his troops and identify enemy locations.

Games DRDO and PSUs Play

Following are some examples:

- DRDO and PSUs resort to outsourcing most of the time. When the military issues a Request for Proposal (RFP), a different one is issued by the DRDO and concerned PSU. Invariably, the military will land up paying greater costs than what they would have paid directly to private vendors.
- CAIR has admitted in writing that it is 'not a developer' responding to a query about security solutions. So, it should be clear that CAIR simply outsources security solutions but retains clout that only they can develop security solutions of 'confidential' and above levels.
- Having secured the consent to develop the ACCS, BEL sent

the proposal to the Defence Procurement Board (DPB). The DPB called the representative of Elbit Systems, Israel (who had supplied the tactical computers for the ACCCS to BEL) separately and asked them whether BEL had done any price negotiation with Elbit. The answer was no. DPB then did the price negotiation and were able to reduce the costs by ₹100 crore, as disclosed by an official who was on the DPB then. The implications are clear and this is just one small example.

- In 2009, one of the Deputy Chief of Army Staff happened to visit Israel. The fielding of ACCS in the Army was to commence later in the year. The Israelis informed the Deputy Chief that the more advanced version of the tactical computer was available for the ACCCS but BEL seemed disinterested. This was despite a clause in the contract with BEL that in case advanced versions of components were available prior to fielding, the latest will be fielded. When both BEL and Elbit representatives were summoned together by Army HQs, BEL agreed to field the latest version of tactical computers. BEL's obvious gameplan was to first field the old hardware and then demand more money to upgrade the same.

Defence Procurement Policy 2013

[8]By opening up to the private sector after decades of being confined to the public, the DPP 2013 comes as a breath of fresh air. Our defence-industrial complex has degenerated to such an extent that as much as 77 per cent of defence equipment is imported, including assault rifles and carbines. It is not that the Army is trying to be fanciful, as some people think. The imports are being used; the Central Reserve Police

---

[8]Katoch, PC, 'Light Beyond the Tunnel', *SP's MAI*, Vol 3, Issue 13, SP Guide Publications, New Delhi.

Force (CRPF) personnel in Jammu and Kashmir can be seen sporting Israeli X-95 carbines. Some time ago, the BSF signed a contract with Italy's Beretta for buying 68,000 sub-machine guns worth more than ₹400 crore. Earlier, the CRPF had signed for purchase of 12,000 X-95 Tavor carbines from Israel costing more than ₹1 lakh a piece. Force One of the Maharashtra Police, which was created post 26/11, is armed with Colt M-4 carbines from the US, Brugger and Thomet sub-machine guns from Switzerland, MP-5 sub-machine guns from Germany and AK-47 variants from eastern Europe. Unfortunately, the indigenous weapons are substandard, forcing the security agencies to import. It should be a matter of shame that we are unable to produce state-of-the-art individual small arms.

Coming back to DPP 2013, it still needs to be refined further to specifically address the requirements of information and communication systems as the procedures are so extended that technology will get outdated by the time these are fielded. The delays are primarily because of the inadequate capacities of DRDO and PSUs as well as the stonewalling of private sector participation in order to buy time. The project to re-energise ASTROIDS has perforce been foreclosed now because ISSA could not deliver the requirement despite delays of so many years. The projects of BSS and CIDSS being handled by BEL are proceeding at an excruciatingly slow pace. The fact is that both the DRDO and PSUs have limited in-house capability and resort to outsourcing to private sector most of the time. Even CAIR, under DRDO, admits it is not a developer. The current stipulation that security solutions of confidential and above classification can be developed only by CAIR is ridiculous considering the agency is actually outsourcing development of such security solutions. The answer lies in making all ICT-intensive projects as 'Make' projects with full private sector participation. Then there is the crying need to simplify and shorten the 'Make Procedure'. It would be prudent for MoD to add such provisions as an addendum to DPP 2013 or at least incorporate them in DPP 2014. Going by past experience, future ICT projects like the TCS, BMS and re-energising ASTROIDS should have 'full' private sector participation. It is equally important for the government to overhaul the DRDO, Ordnance Factory Board (OFB) and PSUs, ushering in true accountability, making them more focused and giving clear directions for what is required to be done by laying down strict timelines. A system of roll-on development, monitoring, periodic checks, feedback and midcourse corrections needs to be instituted. The government must spell out a Weapons, Equipment and Technology Development (WETD) road map on a fifteen-twenty year basis and make R&D allocations accordingly, preferably by grouping DRDO-PSUs and the private sector. This road map must include specific plans and measures to be adopted for leapfrogging technology. The government actually needs to take a call and divide the responsibilities between the private sector and DRDO-PSUs so that defence production gets a focused impetus.

Avoidable Follies

Military Survey: This was brought under DGIS in May 2004 to ensure confluence of the OIS, MIS and GIS in order to accelerate NCW capabilities. However, Military Survey was moved out in 2011 just as they were pushed to produce results, which is a setback to netcentricity. Development of Enterprise GIS has also been stalled because of this, which has had severe adverse fallout on OIS and in achieving NCW capabilities.

BMS F-INSAS Impasse: Phase 3 of F-INSAS (computer subsystem, radio subsystem, software and software integration) which is being developed separately from the BMS is impractical in terms of time and cost.

Project Management Organisations (PMOs): PMOs of DGIS overseeing development of various OIS, running on hard scale authorisation of officers since long, were managing their respective charges with difficulty. Though the concept of the PMO is globally accepted, PMOs in DGIS have been ordered to be scrapped and the manpower reduced. This has resulted in heavy workload for the components. The cells replacing PMOs have even lesser authorisation of officers than what were posted on hard scale in the PMOs. The adverse effects in terms of meeting timelines should be apparent.

Defence Procurement Policy (DPP): DPP 2013 is an opening for the private sector, however, it should have incorporated and shortened and simplified 'Make' procedure and directions for all ICT intensive to be categorised 'Make' with private participation.

Remote Data Policy: Implications of the National Remote Data Policy worked out by the Group of Ministers (GoM) in 2011 are henceforth, OIS will only have one metre resolution. This is a retrograde step as half metre resolution is already available on Google to the terrorists and insurgents and our adversaries would be looking for something even better.

Focus Needed

The government and military need to focus on following issues:

- The defence forces required a C⁴I²SR system 'yesterday' and present and future asymmetric threats and desired responses dictate that this is not a requirement of the defence forces alone. C⁴I²SR must envelop the complete defence and security establishment and concerned government and non government agencies, plus everyone concerned with homeland security.
- The project to evolve common standards and protocols for the Services must be accorded top priority.
- Communication and Information Systems planning should be seamless; horizontally and vertically with adequate safeguards and authority. An integrated communication

network that enables requisite standard signal communication support to all the three Services needs to be established. Even a project like TCS should cater for parallel links with sister Services to meet mission specific requirements.

- The process of automation of all directorates at Army HQs must be speeded up in a time-bound manner. Those that are automated too, are limited to emails only. MoD too, needs to be automated for faster processing of cases. Uniformity of communication and information systems under procurement by the three Services should be ensured including new items like SDRs.

- Evolution of a tri-Service NCW philosophy, tri-Service ICT/communication philosophy, tri-Service cybersecurity and information assurance policy, tri-Service policy for Data Handling and Data Storage should be ensured. Information must be viewed from strategic viewpoint and recognised as mission critical resource.

- The military must evolve and implement an enterprise-level Information Security and Assurance Programme (ISAP). Necessary enablers to provide core competencies for gestating and sustaining the ISAP must be developed as part of capacity building. Presently, numerous applications are coming from all over, some of them without adequate security solutions. This would jeopardise security once total networking is achieved.

- The last mile of Army intranet must be made secure and should be re-extended to HQ IDS. It should also connect the DSSC and CDM. This will enhance jointness and also facilitate e-learning.

- Interoperability within the Services and through the NATGRID needs to be planned. Interoperability between all arms and services of the Army at Battalion/Regiment-level, interoperability with PMF and police for anti-terrorist operations would be the right step in the direction of national netcentricity. NATGRID should eventually be linked to the Global Information Grid.

- Understand that optimising C^4I^2SR by the Services simply cannot be achieved in absence of a CDS. The expanding combined threat from China and Pakistan requires that we optimise and enhance our defence potential post-haste, and a CDS with full operational powers is a vital link in all this.

- Bandwidth requirements for C^4I^2SR need to be viewed keeping in mind the incremental requirements that would be needed progressively over the years. The government needs to examine allotment of a dedicated Defence Band from the spectrum not only to meet the bandwidth requirement of the Services but to guard against future threats to national security.

- Training for the personnel handling the C^4I^2SR needs to be well planned. Aside from individual digitised training packages, the defence forces need to work out comprehensive training packages.

- India still has little indigenous R&D in developing state-of-the-art technologies for networked systems. We must develop our own operating systems, GIS software, computing and networking hardware with standardised proprietary protocols and standards. The industry must go all out in development of new technologies, customised for military use as this is the biggest implementation challenge in our efforts towards netcentricity.

- A level playing field should be ensured which encompasses both the private industries and the DPSUs.

- Robust security algorithms must be speedily developed to ensure security of both stored data as well as transmitted information.
- The MoD should examine how to telescope the long drawn 'Make Procedure'.
- Indigenous capability needs to be developed against enemy Electro Magnetic Pulse (EMP) attacks (nuclear and non-nuclear) as also for checking/testing of hardware and software against embedded malware. CAIR needs to exponentially increase capacity to develop new and varied algorithms in order to keep pace with rapid induction of new systems. The Scientific Advisory Group must find ways and means to accord approvals in telescoped time frame. We need to speedily advance our chip manufacturing capabilities, a sphere in which we are decades behind China and which has serious implications for network and communication security.

The emerging conflict situations call for radical changes in our organisational structures, work culture, warfighting capabilities, doctrines and operational concepts. If India is to graduate beyond a regional power, then our national focus should include establishment of a joint force networked through an effective C^4I^2SR grid. Emphasis must be placed upon indigenous research and self-reliance, to ensure security and redundancy in our system. We must develop netcentricity related concepts and capabilities in models and simulations too. Setting realistic and quantifiable goals, developing an implementation plan to achieve these and in so doing, being able to measure the progress made is essential. An immediate goal must be the availability of a networked joint force as a test bed that can experiment with the concepts and capabilities of netcentricity. Our expertise in information technology must be exploited to create an interface between the defence forces, security establishment, the government and industry with profitable pay-offs for both to foster self-reliance. C^{4}I^{2}SR is a multidisciplinary process. It is, therefore, imperative that some of the critical issues as mentioned are addressed at the inception stage. Evolution of an enterprise architecture, integration and interoperability, communications, bandwidth and latency, introduction of new technologies, robustness of transmission, message and signal routing, sensor exploitation, management of databases, information security, information overload, integrated logistics, dangers of micromanagement, commercial influences, strategic initiatives and time for implementation must all be considered.

6

Military Survey and GIS

[1]Military Survey of the Indian Army and Survey of India (SoI) both owe their origin to the erstwhile British Empire, of which India was a part. Their task was to consolidate the territories of British East India Company and British Empire in India. Though much water has flown under the bridge over the past several decades, Military Survey has apparently not been able to severe the umbilical cord, as indicated by their resistance to adjust and meet twenty-first century requirements. Though the Map Policy of India is explicit that mapping within India is the responsibility of SoI, Military Survey more advertently than inadvertently gets involved internally, rather than focusing on transborder mapping requirements. Even in technology adaptation, SoI has gone way ahead of Military Survey, whereas, it should have logically been the other way around. As a result, the products are primarily Google maps that hardly measure up to military requirements. There appears to be little urgency towards developing and introducing a Geographical Information

---

[1]Survey of India, http://www.surveyofindia.gov.in/

System for the military, a requirement that Military Survey should have met a decade ago.

Presently, Military Survey is thirty years behind in meeting even existing routine mapping requirements of the military, whereas Large Scale Mapping requirements of say 1:5,000 and below is practically not being met at all, which are vital to Operational Information Systems (OIS) being introduced into the military, particularly the Army. [2]It is possible to prepare accurate base map larger than 1:10,000 scale. The development of a specific methodology for preparation of such large-scale map with the use of advanced technologies such as Remote Sensing, Global Positioning System (GPS) and GIS in an integrated way is the need of the hour.

Background History

Ancient Indian history shows that surveys were part of the administration of kings, particularly for revenue collection. While the Mughal period has evidence of its development and application, [3]Raja Jai Singh, during his rule from 1699 to 1743 CE, had established astronomical observatories in Delhi, Jaipur, Mathura, Ujjain and Varanasi, and was known to have keenly studied the European system of surveys. The British realised in early eighteenth century that for them to maintain control over the captured territories, they needed to chart the areas. In India, the East India Company initiated actions to undertake topographical, trigonometric and agricultural surveys, the latter was used to assist revenue collection. Colonel William Lambton, a geographer and a geodesist, is credited with commencing the survey of the country in 1802, having obtained approval of the plan for Geographical and Mathematical Survey

---

[2]Krishnamurthy, J; Natrajan, S; Krishnaiah, P; Kalyanramn, K; Rao, Mukund and Jayaramn, V, 'Global Spatial data Infrastructure: Large Scale Mapping using high resolution satellite data', http://www.gsdidocs.org/gsdiconf/GSDI-7/papers/TSsdSN.pdf

[3]The Jaipur Observatory, http://cadrans-solaires.pagesperso-orange.fr/monde/jaipur/jaipur_uk.html

(later named the [4]Great Trigonometrical Survey by the British) initiated in 1799. In a span of just twelve years (1802-15), Lambton achieved the remarkable feat of covering some 164,342 square miles. When Lambton died in 1823, George Everest succeeded Lambton as the superintendant of the Great Trigonometrical Survey, although he had joined the survey in 1818. He was promoted to surveyor general in 1830 and was succeeded in 1843 by Andrew Waugh. In 1877, Colonel Walker succeeded Andrew Waugh and became the surveyor general. It was under Walker that the branches of trigonometrical, topographical and revenue survey were merged and the Survey of India was established.

As territories of British East India Company expanded, the 'geographical', 'topographical', 'trigonometrical' and 'revenue' surveys got expanded to include 'marine', 'route' and 'military'. Whatever the category of survey, the surveyors accompanied the military wherever they went. It was during the period of Andrew Waugh as surveyor general that heights of some seventy-nine Himalayan peaks were determined and the highest peak named Mt Everest after George Everest. [5]Not many know that the actual measurement of the height of Mt Everest was first done by an Indian mathematician named Radhakant Sikdar from Bengal in 1852 with the help of theodolite and trigonometric equations. He had measured the peak to be at a height of 29,000 feet, the record of which can be found in the SoI museum at Dehradun. At the time, little was known of the Great Himalayas in terms of specific geography and topography, and much of the Great Trigonometric Survey's job was to map this range. At that point in time, Everest was known simply as 'Peak XV'. The peak could well have been named Mt Sikdar.

---

[4]Roy, Rana Deb, The Great Trignometrical Survey of India in Historical Perspective, *Indian Journal of Science*, 21(1): 22-32, 1986.

[5]*Journal of the Asiatic Society of Bengal,* Internet Archive, http://archive.org/stream/journalofasiati291860asia/journalofasiati291860asia_djvu.txt

To start with, Military Survey was formally raised under Military Intelligence Directorate during WWII. It was later brought under the Military Operations Directorate as MO GS GS. Over the years, Military Survey had cells in Military Intelligence Directorate, Engineer-in-Chief's branch, Artillery Directorate and College of Military Engineering and survey branches/cells/sections at Command, Corps and Divisional Headquarters-level. They had a system of reverse deputation with SoI which helped officer management who could be manipulated as required. Military Survey also has a direct link with the Ministry of Science and Technology and annual confidential reports of top-level officers of this organisation are endorsed by this ministry. All manpower of Military Survey below officer rank is from the Corps of Engineers (Bengal Sappers).

When the Indian Peace Keeping Force (IPKF) went into Sri Lanka, they found maps with the Sri Lankan military to be far superior to those provided to Indian Army by Military Survey. After thirty years, the situation probably remains the same. Also, when, the Indian military went into Maldives on the request of the Maldivian President, Military Survey had no maps to offer. Ironically, the director, Research and Analysis Wing (R&AW) could only offer a tourist map. As a result, even the para drop on Malé airfield was planned on a tourist map, which luckily was not required as air landings were eventually possible.

Recent Times

In May 2004, Military Survey was brought under the newly-created Directorate General of Information Systems under express sanction of the Defence Minister to ensure inclusive development and deployment of OIS, Management Information Systems (MIS) and GIS. The need to shift from Platform-centric Operations to Net-centric Operations had brought into focus vital issues. Net-centric Warfare has the critical requirement for integration of operational and tactical information and knowledge with reference to terrain for precise targeting. Battlefield management requires coordination between units, formations, other Services and multiple government agencies. Real time geographical visualisation of the battlefield scenario on a network is required that is possible through exploitation of geospatial data from multiple sensors obtained from space, aerial, ground, subsurface and other platforms. Commonality of data and standards for defence services is an imperative. The tasks assigned to Military Survey include transfrontier mapping, updating maps with satellite imagery, creation of GIS, creation of Digital Topographical Database, preparation of Defence Series Maps (DSMs), Large Scale Mapping, training on GIS and attribute data collection, photogrammetric survey and remote sensing and the like. For Military Survey, the requirement to introduce an Enterprise GIS became paramount, as also did the requirements of Large Scale Mapping in meeting increasing demands of the upcoming OIS. Transfrontier mapping responsibility that was earlier up to 300 kilometre deep across the border (which translates into something like 2,643 topographical sheets of 1:50K scale and 803 topographical sheets of 1:250K scale) was increased to a depth of 5,000 kilometre by Headquarters Integrated Defence Staff (IDS), apparently to meet requirements of the Strategic Forces Command (SFC).

As we move closer towards operating in a fully automated digitised battlefield, there is a need to review the way digital maps and imagery need to be organised and processed for seamless functionality at both strategic and operational-levels. All Battlefield Management Systems (BMS) are inherently dependent on a robust GIS. The GIS applications require geospatial base data in terms of digitised maps, imagery, attributes and so on before it can deliver results to the user. This base data can be considered as the blackboard on which users create their mission specific overlays. In a GIS environment, the user should be able to carry out a number of 2D and 3D analysis functions purely using the base data. Therefore, the quality of the base data in terms of spatial accuracy and the level of attached attribute data, the structure and format of the digitised data and in case of topographical maps, the datum and projections used, assume great significance.

Technology, policies, standards and resources are necessary to acquire, process, store, distribute and improve utilisation of geospatial data for military purposes. Therefore, the need to establish a Defence Spatial Data Infrastructure (DSDI) is never so urgent, especially when a National Spatial Data Infrastructure (NSDI) has already been established albeit its integration with concerned government ministries and agencies is still a long way off. Military Survey had expanded over the years with Centre for Automated Military Survey (CAMS), Army Digital Mapping Centre (ADMC), Defence Institute for Geospatial Management & Training (DIGIT), a Field Survey Group and a Ground Air Survey Liaison (GASL) Platoon, the latter providing aerial cover for survey. However, the organisation had gone somewhat in coma. The system of 'reverse deputation' with SoI had been put on hold for a decade, perhaps on purpose. This created additional two-star vacancies for Corps of Engineers officers in SoI with the added advantage of comfortable peace-time office work. The effect in the Military Survey was equally lucrative, with officers continuously serving from seven-ten years in the department. Such officers, especially in the one-star rank displayed little initiative. The Corps of Engineers did not send young officers to undergo survey training till they had physically reported to Military Survey on posting. When a move was made to restart the process of reverse deputation, SoI intimated that they could only make available officers of major general rank replace colonel-level rank officers in Military Survey. This indicated that the Corps of Engineers had outplayed other sister corps in having a far greater number of major general rank officers by stalemating the reverse deputation gimmick for a protracted number of years.

Functioning Under DGIS

The placement of Military Survey under DGIS in 2004 should have been for 'all purposes' but this unfortunately was not done. The lopsided arrangement was that the Ministry of Science and Technology was in the chain of reporting on the annual confidential report of the officer heading Military Survey and Technical Control was with Engineers Branch. Postings of officers to Military Survey were done by the Military Secretary's Branch in conjunction with the Engineers Branch. Prior to merger with the DGIS, Military Survey was in Military Operations Directorate, where the Engineer-in-Chief's (E-in-C) Branch considered it its 'Fourth Pillar'. The merger of Military Survey with the DGIS brought out a host of shortcomings, some of which were:

- Techniques used for production of maps were archaic.
- Google map downloading was being resorted to as base data.
- There was more focus on own side of the border rather than transborder mapping to meet military requirement despite Map Policy of India categorically stating that all mapping within Indian borders is the responsibility of SoI.
- The organisation and entire focus (without change from the British legacy) was on physical survey within own borders, however, even this had not been done for decades in counter-insurgency areas of J&K and in the North-East despite presence of Army in all such areas.
- Demands for satellite imagery were forwarded to Defence Imagery Processing and Analysis Centre (DIPAC) through Military Intelligence but were never followed up and Defence Series Maps (DSM) were being prepared without incorporating satellite imagery.

- Patrol reports of various Army patrols in difficult areas were ignored for updating maps since this meant hard work.

The Field Survey Group and a Ground Air Survey Liaison (GASL) Platoon of Military Survey placed at Agra, with Air Force providing aerial sorties, basically met requirements of SoI as the aircrafts fly ten kilometres own side of the border.

- While Field Survey Group and Ground Air Survey Liaison (GASL) Platoon of Military Survey placed at Agra and Air Force sorties provided were meant primarily for SoI. SoI was charging money for maps it gave to the Military Survey but in turn was not being charged for the air sorties and the related establishment.
- Field Survey Group and Ground Air Survey Liaison (GASL) Platoon of Military Survey placed at Agra had important maps of even British India period but many are lost as no effort was made to digitise and preserve them.
- Organisation of Military Survey catering for Mobile Field Survey Detachments of Military Survey are to be moved during operations under orders of Military Operations but there were record of many such deployments with Military Survey during 1971 Indo-Pak War, IPKF operations and Kargil conflict.
- QR for posting of officers to Military Survey was civil engineering degree and survey courses but there was no focus on GIS, preparation of maps using IMINT and HUMINT.
- There was neither any policy nor any procedure to gather, update and disseminate terrain data.
- Little/no efforts were being put in towards digitisation of maps, integrating satellite imagery and photography, exploiting advanced technologies and introduction of a GIS.

- No GIS policy and common symbology for the three Services had been evolved.
- There were sustained voids of survey trained officers in Military Survey spanning over a decade.

With the aim of introducing an Enterprise GIS, a tri-Service study was ordered under initiative of the DGIS to examine the nuances for establishing an Enterprise GIS for the military, which was inordinately delayed due to resistance and avoidable delays by the Engineers Branch who claimed that this was their turf. However, once this tri-Service study was concluded, a GIS Policy with Common Symbology for the military was issued in 2009. An RFP to establish an Enterprise GIS was floated by DGIS in mid-March 2009 which has still not seen the light of the day while learning the ropes from the Army, the BSF and CRPF introduced GIS in the years 2008 and 2010 respectively.

More importantly, considering the organisational and output oriented shortcomings of Military Survey (as mentioned above) an Army Study for Reorganisation of Military Survey was ordered, which was headed by a Major General K Surendranath, then Additional Director General, Mechanised Forces at Army HQ, an outstanding officer who later commanded the Test Bed Corps of the Army and finally, the Army Training Command (ARTRAC). Members of the study included representatives of the Military Secretary, Military Intelligence, Engineers Branch, Military Survey, PMO BSS under DGIS, Defence Imagery Processing and Analysis Centre (DIPAC) Naval Intelligence and Air Intelligence. The main issues to be addressed by the study were: Reorganisation of Military Survey units in the backdrop of available global technology and modern techniques; examine existing system of mapping, map updating, digitisation and how updating can be speeded up through reorganisation; examine role of Military Survey in attribute data collection; officer management; rationalisation of existing manpower;

changes in present trade structure; human resource development and present training capability; need for establishing Defence Spatial Data Infrastructure and road map for proposed restructuring. Some of the findings and recommendations of this study were on the following lines:

- Output of Military Survey's archaic mammoth organisation, with an authorised strength of 112 officers, 319 Junior Commissioned Officers (JCOs), 1033 other ranks and 89 civilians, in no way met military requirements.
- DIGIT was functioning on ad hoc basis with limited capability.
- Existing units of Military Survey were functioning on different structures with varying capability, which need to be addressed.
- With same roles, there was considerable duplication of manpower and equipment between the GIS Cells of Military Survey at Command, Corps and Divisional Headquarter-levels (totalling four officers, 78 JCOs and 322 other ranks) and the Image Interpretation Teams (IITs), working under General Staff branch of respective HQ (totaling 98 officers, 145 JCOs and 119 other ranks).

GIS Cells at Formation HQs were no more than storekeepers of printed maps.

- Amalgamation of the GIS and IITs was warranted and would accrue in considerable savings in manpower, equipment and costs besides improving efficiency.
- There was a vital requirement to change Military Survey into an all arms organisation.
- There is a definite need to replace the Military Survey officers at the field formation-level by all arms officers.
- Numerous trades in Military Survey could be reduced to two.

- Military Survey should visualise future operational requirements and cater for infusing new survey equipment and technology.
- Emerging technologies like Digital Photogrammetry using digital aerial photo/High Resolution imagery/UAV inputs, mobile data capture in field using PC tablets, gravity and geomagnetic surveys, Airborne Laser Terrain Mapping (ALTM)/Light Detection & Ranging (LIDAR) survey, online data transfer for updation/web-enabled services, GIS applications and services, Digital Cartography and hi-tech digital planning need to be incorporated.
- Periodicity of updating of maps should be reduced to two-three years instead of the current ten-fifteen years.
- Large scale mapping is a must for future NCW requirements.
- Establishment of the DSDI is an essential operational requirement, which should be preceded by an Enterprise Geographical Information Systems (GIS).
- Part of the savings from the manpower on account of reorganisation could be used to set up DSDI etc.

With the above ongoing study, alarm bells started ringing in the Engineers Branch on account of shrinking turf in utter disregard for operational requirements. Consequently, a hurried conference was organised under the aegis of the then Vice Chief of Army Staff who himself was from the Corps of Engineers. The Engineers Branch put up a case that Military Survey should be moved out from DGIS and placed under Military Operations on the plea that they could manage Military Survey better and oversee their training better since Engineers Branch exercises technical control over Military Survey and gives advice for training, till it was pointed out that Military Survey had just four officers posted at College of Military Engineering (CME) Pune. Besides, in the past five years, no letter of advice on training had been issued to Military Survey by

Engineers Branch and more importantly, not a single officer from this department had bothered to visit Military Survey during this period. On the need to do away with the British legacy of reverse deputation with SoI, Military Survey logic was that this link should be maintained because it can then approach the Ministry of Science and Technology directly (circumventing MoD) for foreign trips—courses, visits/exchange visits etc. The Vice Chief tried to sell the idea of moving DIGIT from its present location in Delhi Cantt to Secunderabad on the plea that DIGIT could interact with civil engineering colleges there. He had to be reminded that there are numerous civil engineering colleges in Delhi-NCR as well. It was pretty apparent that alarm bells had started ringing because of the possibility of:

- Induction of All Arms Cadre officers in Military Survey.
- Military Survey officers in field formations replaced by All Arms Cadre officers.
- Amalgamation of GIS Cells and IITs at formation HQ-level would curtail the Military Survey empire. As it is, the lopsided arrangement continues that the GIS Cell say functioning at Formation HQ-level is not under the GS Branch but continues under the local Engineer unit as this has successfully been evaded by the sappers.

Despite the aforesaid hurdles, the Army Study for Reorganisation of Military Survey headed by Major General (later Lt Gen) K Surendranath, that made major operationally vital reorganisational recommendations, was approved and sent to concerned directorates for implementation. However, Military Survey has managed to freeze it and save some movement on rationalising and reducing the numerous trades in Military Survey.

Senseless Shuttling

As mentioned above, Military Survey was brought under the newly established DGIS in May 2004 under the express sanction and directions of the Defence Minister himself. It was considered vital for operational capacity building; GIS being the domain of Military Survey was required to merge with the DGIS which was already chartered to develop and handle OIS and MIS. Confluence of the OIS, MIS and GIS was designed to accelerate the pace of acquiring NCW capabilities by the Army in particular and the military in general. Military Survey being the provisioning agency for maps to all three Services plus Para Military Forces (PMF) and Central Armed Police Forces (CAPF) on demand from MHA.

What was not known is that the Vice Chief from Engineers before demitting office had apparently extracted a promise from his successor to revert Military Survey back from DGIS to Military Operations for reasons mentioned above. Ironically, his successor issued such orders just before demitting office but not before Military Survey had moved into and occupied a complete floor of the brand new building of DGIS in Delhi Cantonment opposite the Institute for Defence Studies and Analyses (IDSA).

It is well understood that the Vice Chief of Army Staff has the powers to order moves as part of reorganisation of Army HQ. However, what is relevant in this case is that the move of Military Survey in 2004 was under the express sanction of the Defence Minister and reversion should have had the sanction of the MoD, which was deliberately avoided. It is significant to note that ASTROIDS lying defunct in Military Operations for many years eventually had to be moved to DGIS for re-energising. The moving out of Military Survey from DGIS is a retrograde step that will adversely affect the military acquiring NCW capabilities. MoD intervention is certainly warranted.

The Army needs to find a solution to obviate in-Service bureaucracy and give precedence to overall capacity building over consolidation of individual reorganisation. The move of Military Survey to DGIS, though approved by the Defence Minister, was incidentally processed though another Vice Chief of Army Staff who also belonged to the Corps of Engineers (Lt Gen S Pattabhiraman).

Current Scene

- Military Survey has washed off its hands from progressing the GIS/Enterprise GIS on the plea that it is a General Staff (GS) issue. This is despite the fact that it should actually be their bread and butter, they are now part of Military Operations (MO) which is the topmost GS Branch and Military Survey itself is in the old cocoon of MO (GS). Therefore, development of an Enterprise GIS must be the prime responsibility of the Military Survey.

- Presently, Military Survey is occupying a complete floor in the new building of DGIS but is totally unresponsive to the latter.

- Non-establishment of an Enterprise GIS despite the RFP for it in 2009 is adversely affecting development and fielding of OIS. Military Operations located in South Block has little time and inclination to push Military Survey into progressing the Enterprise GIS.

- Military Survey in conjunction Engineers Branch has completely killed the Army Study headed by Major General (later Lt Gen) K Surendranath that was approved in 2009. This is quite apparent because a new study on Reorganisation of Military Survey has been ordered in early June 2013 after waiting for Lt Gen K Surendranath to superannuate from service on 31 May 2013; the least he would have done was to object to the very requirement of a second study.

- Military Survey is back to its leisured engagement of paper maps and continues to be thirty years behind military requirements even in this.

- Ironically, if you physically look across the Indo-Pak border, you can see villages and other construction in some places that do not figure in our maps provided by the Military Survey.

Requirement

The Chinese media reports that the Chinese premier Xi Jinping has cracked the whip on the PLA and police, especially those practising hedonism and engaging in lackadaisical functioning. The MoD and the military need to take a call on shaking up this white elephant of Military Survey that refuses to stir. The following merit serious consideration:

- The 'reverse deputation' of the British era should have actually been replaced by a simple and straight three-four years deputation with SoI, which should have been effected within a few years of Independence. The circumstances under which the British started the process of 'reverse deputation' do not exist anymore. There should be a vertical split between Military Survey and Ministry of Science and Technology, with Military Survey brought under the MoD through Integrated HQ MoD (Army).

- The Army Study on Restructuring of Military Survey which was approved in 2009, should be implemented, Engineer-in-Chief Branch of Army asked for an explanation on why they sat on it and the second study ordered in June 2013 scrapped.

- Not only should Military Survey be made an all arms organisation, it should be headed by a general cadre officer, preferably who has commanded a division.

- Military Survey should be reverted back to DGIS without further loss of time, as was envisaged and approved by the Defence Minister in 2004.

- Development and fielding of an Enterprise GIS and establishment of the DSDI need accelerated focus.

- Vision for Military Survey must include provision of real

> time and accurate geospatial data and services for network-centric applications across the entire spectrum of conflict.

- An in-depth analysis should be done to meet the mapping (including large scale mapping), digitisation, imagery and GIS requirements of the military keeping in mind the fielding of OIS and enhanced missile ranges and a road map worked out to leapfrog large mapping voids.

[6]As we carry out transition from a platform-centric to a network-centric approach, it is essential for us to understand the strengths and limitations of the available technology to harness its full potential. Geospatial data and its exploitation remain central to any battlefield management system, and as such, understanding the complexities of various elements of this type of data becomes imperative. Organising the base geospatial data is in itself a major task. Its further exploitation would remain a major challenge unless all users come to a common framework. Development of any common geographic reference framework is essentially a tri-Service responsibility and can only be undertaken in conjunction with agencies responsible for production of geographic data. In our case too, the three Services are using maps based on differing datums and projections. Even within individual Services, standardisation with regard to maps, imagery and user generated symbology is yet to take place. Issues regarding procedures to be adopted on zone boundaries between the zones of our projected coordinate system have to be worked out. Unless these aspects are addressed collectively by the three Services, results from applications developed for C^4I^2 systems will remain unpredictable, especially during joint operations. It is, therefore, imperative that the issue of common reference framework for geospatial data is addressed on priority. Military Survey has a major role to play in all this and needs to get its act together.

---

[6]Elangovan, *GIS: Fundamentals, Applications and Implementations*, New India Publishing Agency, New Delhi, India, 2006.

7

Geospatial Reference Framework

For an automated Battlefield Management System (BMS), geospatial data is a vital element of information. We are in an era that has graduated from printed topographical maps, naval charts and imagery obtained through wet film cameras to digitised maps and imagery data—having progressed beyond the limited information contained in mere printed maps and imageries. Enhanced data handling capabilities of software applications coupled with the convergence of data has opened up numerous possibilities for optimal exploitation of all information pertaining to terrain analysis and operational planning. Geospatial data plays a crucial role in maintaining the updated situational picture in terms of geographical location of elements of own and enemy forces between all users in the battlefield at all times. In this context, geospatial data is very important to any BMS, considering that its aim is to achieve 'shared awareness' amongst all users.

[1]At the same time, today's dynamic battlefield has diverse users' calls for exchange of positional information between systems using different scales of maps, imagery from different airborne sensors, different map projections and the like. Additionally, latest precision weapons and munitions need much more accuracy for pinpoint targeting in all types of maps, including small and medium scaled ones than what is required for majority of older area weapons. It is, therefore, essential that the precise location of own and enemy forces/targets down to individual soldier/weapon platform are constantly maintained. [2]This highlights the essential requirement that for any BMS to be optimally effective, it must have the requisite common geospatial reference framework.

Geospatial Data

There is a need to review the way digital maps and imagery need to be organised and processed for seamless functionality at both the strategic and operational-levels as moves closer towards operating in a fully automated digitised battlefield. A robust GIS is vital as all BMS are inherently dependent on it. The GIS applications require geospatial base data in terms of digitised maps, imageries, attributes and so on before they can deliver the desired results to the user. This base data can be considered as the blackboard on which users create their mission specific overlays. The user should be able to carry out a number of 2D and 3D analyses functions purely using the base data in a GIS environment. Hence, what assumes importance is the quality of the base data in terms of spatial accuracy and the level of the attached attribute data, the structure and format of

---

[1]'Information Management for Net-Centric Operations', Defence Science Board Summer Study, Department of Defense, Washington DC, USA, 2006, http://www.acq.osd.mil/dsb/reports/ADA467538.pdf

[2]Gould, Michael and Hecht, Louis, Jr, 'OGC: A Framework for Geospatial and Statistical Information Integration', Open GIS Consortium, OGC White Paper, Estonia, September 2001.

the digitised data and in case of topographical maps, the datum and the projections used. All forms of geospatial data use the maps as reference as maps remain its basic element. Though considered as the lowermost layer in a multilayered geospatial data set, topographical maps become the most important element in defining the common geographical location framework for different users in the battlefield. Notwithstanding the fact that we take the maps for granted, transforming the three-dimensional earth on a flat surface in itself is a complex task. Due to the enhancements possible in digitised vector maps by additional layers and the high accuracy levels, other form of geospatial data such as high resolution satellite imagery can augment the map details but cannot replace them. In the case of conventional printed maps, users could reference the maps, carry out corrections and direct other elements in the battlefield using visual observations since an automated system is meant to relay this information seamlessly. [3]However, unless the user at the other end is using the map with the same datum projection, the relayed positional information will be incorrectly plotted on his system, leading to undesired results during execution of operations.

Framework

[4]It is essential to understand the concept of coordinate systems in an environment where positional data is exchanged automatically without user intervention since shortcomings in application design could result in inaccuracies with respect to location information. A spherical coordinate system of longitude and latitude, which measures the point on earth in terms of angular distances between the equator and the poles, represents the spherical earth but since the earth is not a sphere but an ellipsoid, the measurements based on

---

[3]'Understanding Datums and Projections', *Land Information*, New Zealand, http://www.linz. govt.nz/geodetic/find-out/understanding-datums

[4]'What are geographic coordinate systems', *ArcGIS Resources*, February 1998, http://resources. arcgis.com/en/help/main/10.1/index.html#//003r00000006000000

spherical earth will differ significantly with those that use an ellipsoid model of earth. The parameters of this ellipsoid are collectively called 'datum'. In simple terms, a datum is basically a framework that enables defining coordinate systems. In other words, a datum is the mathematical model of the earth used to calculate the coordinates on any survey system. Thus, the Geographical Coordinate System (GCS) takes into account an ellipsoid that best represents the earth. Since the earth is made up of irregular surfaces, calculations for position and direction are different. Hence, cartographers use a mathematical figure (ellipsoid) to overcome these problems. An ellipsoid appropriate for a particular region may not always be appropriate for another region and certainly cannot be applied across the board uniformly. Many such ellipsoids are in use by various countries, and the centre of these ellipsoids may not coincide with the centre of the earth or other ellipsoids in all three axes of X, Y and Z. On balance, a point on the surface of earth will get represented by different longitudes and latitudes depending on the ellipsoid used, its origin and its orientation. For example, the Empire State building in the US has different coordinates in terms of degree, minutes and seconds using National American Datum (NAD) and NAD 83. Hence, it is essential to know the particular earth model that has been used to reference the latitude and longitude of any point on the surface of the earth.

For converting the spherical earth into a flat map, various countries adopt a mathematical ellipsoid that best matches their area of interest. To project the surface on a two-dimensional plane, it is this particular ellipsoid that is used as the reference. However, depending on the type of projection used, certain distortions can occur in the projected map. While the GCS forms the graticules of longitudes and latitudes, to be able to accurately measure distances and calculate directions, the latitudes and longitudes are converted to Cartesian coordinates (X, Y). This forms the Projected Coordinate

System and is referenced as Easting and Northing. The selected ellipsoid takes care of the horizontal datum (X, Y) but another datum called Vertical Datum is used to reference the base from which the elevations are measured. The Mean Sea Level (MSL) is generally taken as the base for vertical datum in most local datum but since the mass of the earth is not uniform, the MSL too, differs from place to place. Resultantly, the true location for any point is closely linked to the horizontal and the vertical datum.

[5]Though there has been considerable excitement concerning the World Geodetic System (WGS) reference framework, it must be understood that as the global datum, it is yet to mature. The system was developed from a worldwide distribution of terrestrial gravity measurements and geodetic satellite observations. The WGS primarily references an ellipsoid whose placement, orientation and dimensions best fit the earth's equipotential surface that coincides with the geoid. Several different ellipsoids have been used in conjunction with the WGS ellipsoid. It takes into account the geoid formed by the earth due to uneven distribution of mass and resultant differences in gravity. Success of this model is dependent on the gravity readings being made available by various countries. For calculating any position on the earth, the US satellite based navigation system uses the WGS reference framework. Instead of using the MSL, the framework uses the Height Above Ellipsoid (HAE) reference for elevations. Significantly, the difference between the two could amount to anything up to 30 metres at places. Whether the earth model used is spherical or referenced as ellipsoid, the geographic coordinates so created have to be converted to Cartesian coordinates as (X, Y) and measured in metres or the distance from the origin (0, 0). A map projection basically projects the graticules

---

[5]World Geodetic System 1984, http://www.oosa.unvienna.org/pdf/icg/2012/template/WGS_84.pdf

found on an ellipsoid on to a plane surface. Subsequently, grids are applied to these projections to provide a rectangular system for referencing and measurements. This is achieved by defining a relationship between the grid and graticule. The projections are primarily of three types: Cylindrical, planar and conical. Depending on the projection used, certain deformation will take place. No single projection is free of deformation.

The Indian military uses the Everest datum, conical projection (polyconic) and Lambert Conical Conformal (LCC) for the military grid. These grids are defined by origin and false Easting amongst other parameters to ensure positive values.

Battlefield Management

It needs little emphasis to say that a common geographic reference framework is essential to retain a common situational picture at all times. The US has faced problems related to this issue both in tri-Services joint operations and while operating with allies. Services of the US military are using different systems of calculations. The US Army uses the Universal Transverse Mercator (UTM) grid and its own Military Grid Reference System (MGRS), the US Navy uses geographic coordinates in degree and minutes whereas the US Air Force uses geographic coordinates calculated in decimal degrees. Add to this the aspect of datum, with the Army using at times the local datum while the other two Services use the WGS 84 datum. [6]There have been instances during operations in Bosnia, Iraq and Afghanistan wherein the datum being used by the three Services ranged from local to European to the WGS Reference Framework. As a result, fire from naval guns was off the intended target by as much as 750 metres. This has important lessons for the Indian military, in that, it is preferable that all elements of any joint

---

[6]'DoD Military Lessons Learned—Joint Army, Air Force, Navy, Marines', http://www.au.af. mil/au/awc/awcgate/awc-lesn.htm

operation should adopt the common reference framework and in case that is not possible, transformation of projection, datum or both should be applied. This would require applications to be developed to take care of these aspects which are transparent to the user. However, the second method (transformation) needs to be viewed with caution since different methods of transformation would yield different results.

Another aspect related to maps pertains to targeting information. The ability of the GIS applications to carry out unlimited zoom on the digitised maps leads to misconception that the accuracy is increasing. A map is accurate only to the extent specified for a particular scale. Any zooming of the map will not result in corresponding increase in details. Precise targeting makes use of observations from multiple sources including human intelligence, meteorological conditions, satellite/aerial imagery and so on. The precision munitions in use rely on WGS 84 datum for navigation, and employment of these weapon systems involves using high resolution imagery of the target referenced to the map of the area for getting the coordinate information. Unless the users of an automated BMS adopt a common framework for handling the geospatial data, there are bound to be errors regarding positional information. This framework should not be restricted to the map datum, projections and grid coordinate system but should also deal with the aspect pertaining to file formats, coordinate exchange formats, data structure for all vector maps, common symbology, standard attributed, raster formats and raster file sizes.

Exchanges

While maintaining the positional accuracy, a common geographic reference framework is essential for exchange of geospatial information. While digitised photographic maps form the lowest layer of data, each user generates certain amount of specific data over these maps as overlays. This data could include vector layers as well as raster imagery obtained from UAV or other sensors. However, it must be understood that unless the data is made to conform to the common reference framework, its interpretation by other users will lead to errors. For example, a feature on a satellite imagery projected for [7]UTM using WGS 84 framework would not reflect the same coordinates when placed in a coordinate system using [8]Everest Datum and [9]LCC Projection. Any re-projection of satellite imagery also tends to result in certain distortions and loss of information during re-sampling. Similarly, any symbol, line or polygon drawn on an overlay picks out the coordinates from the maps which are used as the base. Their use in applications using a different coordinate system and datum would give varying results.

In the context of joint operations, it is essential that the three Services adopt a common geographic reference framework that best meets our requirements without involving a major change in the way we have been handling maps. This framework could also be expanded to include raster data like satellite imagery, air photographs etc. in terms of image registration, resolutions, attached meta data and resolution (spatial, spectral and radiometric). The framework should also define a standard for referencing the imagery data corresponding to the topographical data in terms of image tiles for easier exploitation of imagery. Indexing of imagery data in a manner similar to that of digitised maps would ensure better management, exploitation and commonality of base data used by the three Services.

---

[7] Riesterer, Jim, 'UTM - Universal Transverse Mercator Geographic Coordinate System', http://geology.isu.edu/geostac/Field_Exercise/topomaps/utm.htm

[8] 'Coordinate transformation from WGS 1984 to Everest 1830 or vice versa', *ArcGIS Resources*, http://forums.arcgis.com/threads/53532-Coordinate-transformation-from-WGS-1984-to-Everest-1830-or-vice-versa

[9] Lambert Conformal Conic, http://www.georeference.org/doc/lambert_conformal_conic.htm

Essential Requirement

While the Indian military makes the transition from platform-centric to NCW capabilities, it is essential for us to understand the strengths and limitations of the available technology to harness its full potential. The HQ IDS, DGIS of the Army and Navy/Air Force sister directorates and the Military Survey all have a major role to play in all this, especially since Military Survey has moved out of the ambit of the DGIS. Geospatial data and its exploitation remain central to any BMS and as such understanding the complexities of various elements of this type of data becomes imperative.

Organising the base geospatial data is in itself a major task. Its further exploitation would remain a major challenge unless all users come to a common framework. Development of any common geographic reference framework is essentially a tri-Service responsibility and can only be undertaken in conjunction with agencies responsible for production of geospatial data. In our case too, the three Services are using maps based on different datums and projections. The nuances of joint Services operation scenarios must be fully visualised and catered for. Even within individual Services, standardisation with regard to maps, imagery and user generated symbology is essential and not fully streamlined yet. Issues regarding procedures to be adopted on zone boundaries between the zones of our projected coordinate system have to be worked out. Unless these aspects are addressed collectively by the three Services, results from application developed for C^4I^2 systems will remain unpredictable, especially during joint operations. It is, therefore, imperative that the issue of common reference framework for geospatial data is addressed on priority.

8

Geospatial Intelligence

The security, economic progress and well-being of the citizens of a country are closely interlinked with its intelligence gathering capability. Intelligence plays a prominent role in the decision-making process of any organisation including the military and as such, [1]Geospatial Intelligence is vital to the Command, Control, Communications, Computers, Information and Intelligence (C^4I^2) system. The technological revolution, particularly in the field of IT, and the challenges of the twenty-first century to security, including a growing threat of international terrorism, has forced governments the world over to adopt organisational restructuring. The emphasis is on state-to-state cooperation and exchange of information rather than to pursue an independent agenda since these threats are transnational, omnipresent and can occur at any time. The geographic database is vast; it is no more restricted to areas adjoining international boundaries, and also includes own territory. Therefore, the amount of

---

[1]'Geospatial Intelligence in Joint Operations', *Joint Publication* 2-03, US Army, 31 October 2012, http://www.dtic.mil/doctrine/new_pubs/jp2_03.pdf

information that needs to be processed to filter meaningful intelligence is colossal. This makes exploitation of technology essential in order to facilitate real time intelligence. Conventional approach will not do since it will be time and resource consuming. [2]Automated Decision Support Systems (DSS) can process and form a cohesive picture from large amount of data in fairly short span of time, thereby permitting retention of initiative. The aim of any such system should, therefore, be to fuse data from multiple and sometimes even, unrelated sources to be able to produce a comprehensive intelligence picture dealing with all aspects of information.

Geospatial Intelligence

There are many definitions for Geospatial Intelligence. However, the most apt definition depicting the magnitude of the effort required to build a database of Geospatial Intelligence is by the US National Imagery and Mapping Agency (NIMA) which defines it as 'The exploitation and analysis of imagery and geospatial information to describe, assess and visually depict physical features and geographically referenced activities on the earth.' This elucidates the type of database required by government agencies entrusted with the task of national security. The need for commanders at all levels to have access to maximum data pertaining to the topography and demography of an area of interest is well established. Transparent overlays with supplementary information placed over printed topography sheets have been in use for long and are just one example of this necessity. Computers and associated technology have ushered in the era of digital mapping where this supplementary data can be tagged to various terrain features, which can be displayed, queried or processed based on requirement, without cluttering the base map.

---

[2]Power, DJ, 'What is a DSS?', *The On-Line Executive Journal for Data-Intensive Decision Support 1(3)*, 1996, http://dssresources.com/papers/dssarticles.html

The capability and capacity of any application based on a GIS is dependent on the quantity and quality of data provided as input. A digitised cartography map needs enormous volume of attribute data before it can become suitable as a GIS input. While generally this data has been assumed to pertain to terrain feature only, a true GIS-ready map should logically cover the complete spectrum of data required by various disciplines of geodetic science that has military value in assisting operations. This definitely calls for inputs even from agencies other than those dealing with topographic survey in order to make it all-encompassing. For example, it should assist in calculating the effects of certain type of munitions like incendiary, chemical or biological that cannot be calculated purely by use of standard two-dimensional models, since it also requires detailed knowledge of terrain in terms of type of forest, condition of the shrubs and metrological conditions.

Infrastructure

To reinforce the above, the security forces will need to rely on all relevant agencies of the government in order to optimally exploit the available information in a geographic context. Stand-alone reliance can leave large gaps in intelligence which would be crucial at the time of critical decision-making. It is, therefore, essential that any agency entrusted with the task of producing Geospatial Intelligence should be multifaceted with representation from all agencies dealing with aspects of intelligence, cartography, oceanography, civic infrastructure creation, public utility/safety services and so on. Cartography in an urban environment is a challenging task since there is lack of natural features that can be selected as control points plus the problem of line of sight for surveying. The requirements of users, especially security forces, are also very demanding in an urban set-up. The layout of the sewage system, power and telephone lines/distributors, other communication infrastructure, traffic conditions at various times of the day, location of medical and health centres, the demography, location of trouble spots, ethnic/communal break up of population, architectural drawings of the buildings etc. are the disparate sources of information which may enable a commander in taking logical decisions in case of a crisis.

US/NATO Scene

NATO forces operate in areas other than their own countries; geospatial coverage is focused on area of operation other than homeland. The US, post 9/11, undertook major restructuring of government agencies and formed National Geospatial Intelligence Agency (NGA) by merging in it elements of Central Intelligence Agency (CIA), Defence Intelligence Agency, National Reconnaissance Office (NRO), State Department etc. [3]The NGA's mission statement is 'Provide accurate and timely Geodetic, Geophysical, Geotechnical analysis and Geospatial Intelligence information to support national security, Department of Defence and intelligence objectives' and it is undoubtedly a major arm of the US government, assisting in implementing the national goals of the US. Within the US, the US Geological Survey is responsible for survey and mapping, base data at scale of 1:24K, which is available on the internet. Data for operational areas on foreign soil is prepared by agencies like NGA/Army Geospatial Centre (AGC). More importantly, geospatial data is provided by a single source (NGA-AGC) and interoperability pan military is ensured through common standards, protocols and Application Programming Interfaces (APIs) like NVG, MOLE, C-JMTK, VMF etc. Other major nations too, are restructuring to establish agencies dealing with this important aspect of intelligence. One may argue that the US is a global power and requires a much larger database of spatial intelligence to support its foreign policies and its military, deployed in various parts of the globe.

---

[3]National Geospatial-Intelligence Agency, https://www1.nga.mil/Pages/default.aspx

While we may not draw comparisons with the US in terms of infrastructure at this stage, we will surely have to reach a bare minimum level to match our status as an emerging regional power, including in the Indian Ocean Region. For production and analysis of data required as base for Geospatial Intelligence, numerous agencies are involved. This would include high resolution imagery products, elevation data through ground/air/space based survey, ELINT and HUMINT assets, cartography data, inputs from agencies dealing with internal security, urban and rural development agencies, forest management agencies, intelligence community and so on. It is only when the seemingly unrelated events collected individually by these agencies are merged that a comprehensive intelligence picture in the geographic domain can emerge.

GEO Intelligence in Afghanistan

[4]In Afghanistan, the base data is provided by the US NGA which is supplemented by the AGC. Imagery and elevation are through the BuckEye Program; colour imagery (48 megapixel) at 10-15 centimetres resolution and LIDAR based elevation data with one metre resolution. Situational updates like IED incidents and deployments are forwarded by users up the chain for consolidation at regional HQs. GEOINT cells are located at each HQ. [5]Engineer detachments use ENFIRE to collect data. Collated data from all sources is used to refine digital maps and issued to units at a frequency of every three months.

Geospatial Intelligence comprises imagery, imagery intelligence (IMINT) and geospatial information. The full utility of Geospatial

---

[4]'Geospatial Defense and Intelligence in Afghanistan', 6th Annual Geospatial Intelligence Middle East 2013, GIS for Defense, Intelligence & Security, http://www.geospatialdefence.com/redForms.aspx?eventid=7079&id=389080&FormID=%2011&frmType=Additional%20Content&m=4888&FrmBypass=False&mLoc=U&SponsorOpt=False

[5]'Instrument Set, Reconnaissance and Surveying (ENFIRE)', Northrop Grumman Corporation, http://www.northropgrumman.com/Capabilities/ENFIRE/Documents/ENFIRE.pdf

Intelligence comes from the integration of all three, which results in more comprehensive, tailored Geospatial Intelligence products for meeting the wider requirements of the military, especially in execution of special operations. Sub-conventional and asymmetric conflict scenarios have heightened and the need for special operations is likely to expand in the foreseeable future. Geospatial Intelligence supports special operations, giving them the ability to rapidly respond to threats providing geo-referenced visual and data products that serve as foundation and common frame of reference. These products include interactive maps, virtual fly-through and walk-through mission scenarios.

Indian Scene

On the Indian front, establishment of the National Spatial Data Infrastructure (NSDI) has been a good initiative but its networking with concerned government agencies relevant to geospatial database and updates is still catching up. More importantly, NSDI deals with only some aspects pertaining to creation of metadata of available geospatial data and does not cater for inputs from intelligence community or for that matter the defence services. In spite of the recommendations of the Kargil Review Committee as well as recommendation of GoP on issues related to streamlining the intelligence agencies in the country, a lot of work is yet to be done on issues related to fusion of intelligence data. While the Defence Intelligence Agency (DIA) and National Technical Research Organisation (NTRO) have been established but such steps are still not sufficient to deal with aspects pertaining to Geospatial Intelligence.

As for accuracy and updating maps within India by the SoI, the story appears very poor. [6]After carefully studying Google Earth

---

[6]'Indian security forces discover 'unknown hill' using g-tech', *Geospatial World*, 23 April 2012, http://www.geospatialworld.net/News/View.aspx?ID=24574_Article

satellite imagery, a CRPF task force zoomed in on couple of structures they identified as a 'Naxal camp' in Abujhmad jungle (which straddles the border of Chhattisgarh and Maharashtra) in April 2012. However, when they arrived in the area, the 15-20 thatched huts seen for the first time on Google Earth were actually homes of Muria tribals. S Elango, DIG (Operations), CRPF, later informed the media that nobody knew there was a village called Bodiguda. This village was discovered for the first time since India's independence. The tribals of the village, who have never seen or heard of electricity, water taps, schools, dispensaries, men or machines, have grown up believing the Naxals to be the government. Behramgarh, a village 29 kilometres away, is the closest thing to civilisation.

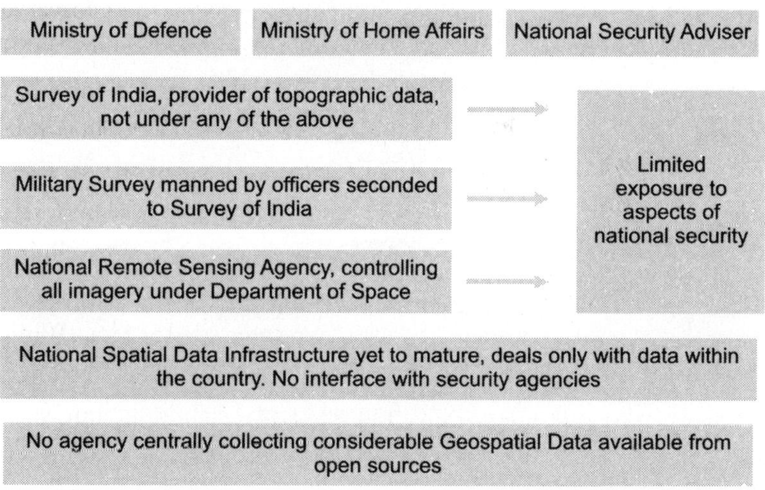

The DIA is the central repository for all intelligence inputs pertaining to the three Services including Imagery Intelligence (IMINT) and Electronic Intelligence (ELINT). However, we are yet to integrate the aspects of topography with the DIA. Within the existing set-up, adequate resources in terms of remote sensing, ELINT payloads and cartography are not available to produce high-quality fused data. Similarly, much more is required at the national-level in terms of integration of various government agencies. While certain isolated linkages between certain government agencies are already in place, this connectivity has to be extended to all necessary arms of the government over a national security information grid for optimal exploitation of various multilayered data sets. Ideally, a comprehensive Geospatial Intelligence data set should be able to generate large-scale maps, surface models to include natural and man-made structures, walk-throughs of critical infrastructure/buildings, computer models to predict and manage natural disasters and many other functions required to support aspects pertaining to nation building as well as national security.

The Disaster Management Authority (DMA) by itself is recognition of the importance of geospatial science in nation building. An organisation like this would have to deal with varied amount of data pertaining to disparate sources of information to perform its tasks. A NATGRID is a basic prerequisite for the collection and assimilation of data coming from different parts of the country, dealing with different aspects of national security. Our existing communication infrastructure is definitely not in a position to deal with information flow of the magnitude necessitated by disasters like the tsunami and recent calamities in Leh, Sikkim and West Bengal. Our NATGRID is already under establishment, which needs to be speeded up by synergising 'all source' intelligence and duly linked to the NSDI. Be it disaster management, communal riots or counter-insurgency operations, the disparate information held by different agencies has to be fused to form an intelligent picture before operations can be planned to produce maximum results. Within the military, the primary prerequisite would be the fusion of inputs from the three Services pertaining to the defence aspect of the national security apparatus. While it will take time to influence smoother flow of information from other arms of the government, the military itself must focus on integration of various sources of information belonging to the three Services under the aegis of HQ IDS.

In the absence of a dedicated defence space programme, the military is dependent on the National Remote Sensing Agency (NRSA) and commercial satellite-based imagery for some of the IMINT requirements. There is a need to employ other alternate sensors to meet defence requirements not only pertaining to imagery, but also to the data required for cartography. This calls for a well-coordinated surveillance effort by airborne sensors of the three Services. Similarly, the HUMINT and ELINT data needs to be fused with IMINT and the topographic data. Military Survey needs a large amount of data to prepare accurate maps of areas across the Indo-Bangladesh/Line of Actual Control/Line of Control. Unless we can task our assets according to the overall vision for Geospatial Intelligence, we will continue to operate in isolation, and remain critically short in terms of geospatial data. There is a need to chalk out a well-thought-out road map in this context.

National Map Policy 2005

[7]Our National Map Policy 2005 defines two series of incompatible maps: Defence Series Maps (DSM) based on WGS 84/LCC and Open Series Maps (OSM) based on WGS 84/UTM. There is no mention anywhere of the elevation system to be used—whether it should be WGS 84 or another. Moreover, the policy does not cover the nautical and aeronautical charts. The policy is restricted to small scale maps and is silent on responsibility for attribute collection. The overall implications are that we have two incompatible projections and associated different grids will be an operational nightmare.

Remote Sensing Data Policy 2011

[8]The National Remote Sensing Centre is vested with the authority to acquire and disseminate remote sensing data. All data of resolutions up to one metre is distributed on a non-discriminatory basis. All data better than one metre resolution is to be screened and cleared by appropriate agency prior to distribution. This policy also talks of specific sales/non-disclosure agreements for data better than one metre resolution. The implications of this policy are that it places undue restrictions on genuine users for the simple reason that point five resolution data is available in the public domain through Google Earth.

Own Capability to Exploit Spatial Data

The fact is that much better spatial data is available in public domain and by extension to our adversaries. Then, applications used by security forces are restricted from connecting to the internet.

---

[7]'India's National Map Policy: HOPE vs HYPE', *Coordinates*, August 2005, http://mycoordinates.org/india%E2%80%99s-national-map-policy-hope-vs-hype/all/1/

[8]Remote Sensing Data Policy (RSDP – 2011), http://www.isro.org/news/pdf/RSDP-2011.pdf

There is undue security concern on base geospatial data. There is no mechanism to exchange geospatial data amongst the various agencies. We have separate agencies for topographic data and remote sensing with no data interfaces between the two. Finally, there is no clear area of interest defined or agencies tasked to prepare and update spatial data. A realistic assessment in case of the military's capability in this respect can be summarised as under:

- Geospatial data available in public domain is of far better quality than that made available to own troops.
- DIPAC and MI-17 (Military Intelligence) deal with IMINT focused on spot imagery.
- There is inadequate high resolution data capability and reluctance to acquire commercial imagery.
- There are severe restrictions on procurement of commercially available satellite imagery.
- There is no capability of attribute collection in areas of operational interest.
- There are limited airborne sensors and many restrictions on their employment.
- Data hoarding and reluctance to network are prevalent.
- Security paranoia even over base geospatial data is rampant.

Required Focus

The following focus is required by the government and the military, as relevant:

- India is the only country with two projection systems—UTM and LCC. This needs to be corrected forthwith and the Map Policy redefined by addressing flaws as mentioned above.
- Unless the National Map Policy and Remote Sensing Data Policy converge and an agency is identified to deal with geospatial data in reality, neither the NSDI nor DSDI can become a reality.

- There must be clear division of mapping—Military Survey for transborder mapping and SoI for mapping within the country.
- Military Survey and SoI are manned by personnel with little expertise in national security and more significantly both of them are neither under MoD or MHA. This needs correction. Military Survey should be placed under MoD through HQ IDS. Similarly, SoI should be directly under MHA. Links of Military Survey and SoI with Ministry of Science and Technology should be through MoD and MHA respectively.
- Data links of NSDI with concerned government agencies (including intelligence agencies like NTRO and the military) need to be established on priority.
- The military should focus on integrating topographical aspects with the DIA. The need to establish a DSDI was never more, the initiative for which needs to be taken by HQ IDS.

Efforts pertaining to geospatial data and intelligence within the three Services need to be integrated to help field a fully functional C^4I^2 system, which would be an integral part of our war-making effort in the years to come. Individual Service approach cannot suffice. No automated BMS can be fully exploited unless quality data is provided to it as input. Our efforts to enable production of Geospatial Intelligence would meet most of the data requirements of both C^4I^2 as well as $TacC^3I$ systems of the three Services. The need to bring intelligence and geospatial information under the aegis of one single agency cannot be underestimated and should not be relegated to a later day. India must reach a base level to match our status as a regional power. In times to come, as India gets increasingly involved in UN backed conflict management activities, we will have to continuously review our policies and carry out organisational adaptations to operate seamlessly with other major member countries. [9] A holistic view of our organisation in the field of intelligence and Geospatial Intelligence is necessary to lay down a framework for changing our national security apparatus to meet future challenges.

---

[9] Katoch, PC, 'Transform National security Apparatus', SP's LandForces.net, http://www.spslandforces.net/story.asp?id=140

9

Information Assurance

The military is making concerted efforts to modernise. As part of this drive for transformation, enterprise information system assets and initiatives are a key focus area. At the tri-Services level, the Defence Information Assurance & Research Agency (DIARA) at IDS is charged with cybersecurity and information assurance of the military. Then are the individual cybersecurity establishments of the Army, Navy and Air Force, where the emphasis is more on cybersecurity than on information assurance. But the Navy and Air Force being smaller Services and with higher levels of intra-service netcentricity are better organised. Therefore, in this chapter, discussion will be more about efforts towards information assurance at tri-Services and Army-level.

Within the Army, the Directorate General of Information Systems (DGIS) ensures that fielding of enterprise-level integrated infomation systems is carried out in line with the laid out doctrine as well as a strategy that leverages integration as the key enabler for transforming collaboration amongst people, organisational processes and technology platforms into a cohesive and synergised

Information Age warfighting capability. A vital challenge in this endeavour is the attempt to transform a legacy operational framework into an Information Age e-enterprise. In an e-enterprise, the tight coupling between functional processes and the underlying information infrastructure amplifies the effect of hardware and software security failures simply because security breach in any single entity whether hardware or application can jeopardise security of the entire network. This signifies the need for security management of not only the information infrastructure but the information itself, thereby necessitating comprehensive information assurance measures to ensure that this mission critical resource is available even in most adverse times.

Our Supply Chain Vulnerability

Almost 95 per cent computer parts come from China. Even telecommunication equipment, including 90 per cent of the state-owned Bharat Sanchar Nigam Ltd (BSNL), is Chinese. We have little capability in Integrated Circuit (IC) chips and don't even produce items like pen drives. On top of that, we do not have any capability to check imported products for embedded malware, bugs, Trojans etc. These vulnerabilities seriously affect the nation and the military. China has a very strong manufacturing base for ICT products. It also has a full spectrum ability encompassing design, development, production and upgradation of IC chips. Besides being a leader in 'fables' manufacturing, China has a wide base of FABs (IC fabrication facilities) spread across the country. Its ICT production companies like ZTE and Huawei bid for international projects. There is a distinct possibility of this prowess being used as a distinct advantage in cyber warfare by way of both Computer Network Attack (CNA) and Computer Network Exploitation (CNE). This is of particular significance since India imports bulk of ICs and systems, thus introducing vulnerabilities in the system which can be exploited either for disruption and destruction. [1]Supply Chain Risk Management (SCRM), therefore, assumes great importance for protection of our critical infrastructure and systems. In a globalised environment, widely dispersed supply chain of key components may provide an intruder with opportunities to manipulate those components or penetrate the distribution chain with counterfeit products. Numerous studies of the vulnerable microelectronics supply chain have identified potential problems with both 'trust verification' and 'uninterrupted availability' of supplies. Adopting measures for both upstream and downstream protection against supply chain attacks would be an essential capability feature for cyber warfare. Also, immediate measures are needed to overcome this strategic deficiency in Indian context by setting up FABs and 'Fabless' manufacturing facilities.

Computer Network Operations

CNO essentially consist of Computer Network Attack (CNA), Computer Network Defence (CND) and Computer Network Exploitation (CNE). It will be fair to assume that CNE for espionage and collection of useful data will be conducted on 24/7 basis even during peacetime. CNA will be launched much before the commencement of active hostilities with the aim to interfere with the command and control and logistics networks. CNA activities are likely to continue much after the cessation of hostilities.

[2]China is expected to launch active Psychological Operations and Perception Management efforts through social media like TV, radio and mobile phones as widely reported. Such an attack

---

[1] Charney, Scott and Werne, Eric T, 'Cyber Supply Chain Risk Management: Toward a Global Vision of Transparency and Trust', Microsoft Corporation, 26 July 2011, http://www.microsoft.com/en-us/download/confirmation.aspx?id=26826

[2] 'Chinese Military Modernisation and Force Development: A Western Perspective', Centre for Strategic & International Studies, USA, http://csis.org/publication/chinese-military-modernization-and-force-development-0

will exploit the religious, communal, caste and diverse development of India. A well-coordinated and deliberated social engineering attack coupled with a few physical ones causing considerable collateral damage could prove to be India's Achilles' heel. In 2012, breakdown of western and northern India's power grids; power blackout at Terminal 3 of New Delhi's airport; frequent blinking of air traffic control radar in New Delhi; a failure of the North-East power grid; frequent penetration of our networks, both for collection of intelligence and to cause disruption, are a few of the distinct possibilities of CNA and CNE during peacetime. It just signifies that such *events* could be triggered by cyber intervention by state or non-state players. Some CNE operations may be designed purely for reconnaissance purposes to map our network topologies and understand their relationship with the command and control structure. Such reconnaissance conducted during peacetime can support offensive operations during war. CNE can also be used during peacetime to install malware which can provide intelligence during this period and later can be triggered during war. The Chinese operators could create intentional 'noise' on the networks that elicit Indian reaction, thus enabling the attackers to gather intelligence on what would be the Indian defence's response under such circumstances.

Relevance

As part of the modernisation drive including acquiring NCW capabilities, the military, particularly the Indian Army has a number of information systems in various stages. It has also felt the necessity to adopt an enterprise-level view of all these systems which are either deployed, being deployed or are proposed to be deployed in the future. This has necessitated the need to adopt an overarching perspective for the vital aspect of information security, not only in the context of environment of the information systems but also the information contained within the systems and their physical assets.

Information Security

[3]For an information system and the information it contains, certain control objectives provide the foundation for all derivative security requirements. These extremely important security control objectives are: First, Confidentiality. At times called secrecy or privacy, confidentiality is the protection of the assets of an information system so that the assets are accessible only to authorised parties, an example is the protection of privileged information in transit through the information infrastructure with the use of encryption technologies. Second, Authentication. It establishes the validity of a user's identity, and is fundamentally verified and proven identification of the user to the information system. It determines that a user is identified to the information system and authorised the use of that particular asset of the information system. For instance, validation of identity before acceptance of commands or prior to release of information. Third, Availability. This ensures that the information and system resources are accessible to authorised system users. A user of the system can be a person or another system that requires a given asset, and as an example, information contained on enterprise servers that is available to mobile users at remote locations. Fourth, Integrity. This implies that information assets can be modified only by authorised parties in authorised ways. For example, a user may be authorised to read data from a database but may not modify that information. Fifth, Non-repudiation. It is proof of the identity of both the sender and the recipient, preserving the responsibility of both the originator and recipient for their actions. An example is validation of a user's signature by the information owner and validation of the information owner by the user prior to acceptance of a transaction.

---

[3]Plain English ISO IEC 27002 2005: Information Security Control Objectives, http://www.praxiom.com/iso-17799-objectives.htm

Information Assurance

Delivery of information in real/near real time is of extreme importance. To have information which is perfectly secure but delivered too late to take operationally critical actions does not fulfill the mission. Hence, information assurance control objectives are required to complement security control objectives and provide the basis for subsequent reasoning about system feature reliability. Information assurance refers to the fact that information is made available to authorised users, when requested, with expected integrity. It encompasses not only the basic system security properties, but also policies, procedures and personnel that are used to maintain the system and its data in a known and secure state. [4]Information Assurance objectives that support security control objectives are: One, Personnel Management. It refers to the personnel practices that support the administration of the security functions of the information system. Two, Vulnerability Management. It refers to the maintenance of the info system software update process to ensure all known vulnerabilities are corrected, i.e. the employment of a comprehensive vulnerability scanning and remediation capability. Three, Configuration Management. It refers to that part of the information assurance system that tracks the hardware, software and firmware configuration of each physical device and allows the info system to be maintained in a known, secure state at all time. Four, Secure Software Development Management. It is the systemic use of software design principles and processes through a Security Development Lifecycle to ensure that the information system software is secure by design. Five, Verification Management. It is the process of testing and validation used to ensure that the system

---

[4]'Information Assurance for Allied Communications and Information Systems', Combined Communications Electronics Board (CCEB), ACP (F), NATO, December 2008, http://jcs. dtic.mil/j6/cceb/acps/acp122/ACP122F.pdf

works correctly, i.e. the maintenance of an independent verification and validation programme which includes, at a minimum, unit, subsystem and system verification procedures.

Present Capability

DIARA: Roles envisaged for DIARA include cybersecurity, risk assessment, disaster management, cyber forensics for the military and training of MoD's cybersecurity personnel. Also, it is required to be the central repository of cybersecurity and information warfare tools for the military. The primary responsibilities include conducting internal/external vulnerability and penetration testing, critical asset evaluation, threat identification and risk assessment. However, limited forensic analysis capability has been acquired. A testing and evaluation laboratory is established for evaluation of security products and solutions. But classification as a certifying authority for the Army has hit a dead end with non-availability of specialists (scientists, mathematicians) and a stipulation that no agency below the Scientific Analysis Group was permitted to certify products of confidential and above classification status. In China, the PLA is spearheading the massive cyber warfare programme. However, there is practically no offensive content to the Indian military's cyber warfare programme.

Army: The Army Cyber Group (ACG) was established eight years ago as Army Cyber Security Establishment (ACSE) to enhance security of the Army's information infrastructure through proactive actions and collaboration. The ACG issued a cybersecurity policy in 2007 and various guidelines for desktop users and for audit and initiating IT projects. Advisories too, are issued regularly and audit security reviews of information bearing networks and systems are undertaken. Cert-Army has been established with a website on the Army intranet. Advanced skills are being sought for vulnerability analysis of networks. Limited forensic analysis capability has been acquired. ACG is also engaged in spreading security awareness in the Army and vetting of IT projects.

Related existing capabilities to the information assurance control objectives clearly show that only issues relating to Personnel Management and Vulnerability Management are being addressed and that too, in limited form. The other information assurance objectives of Configuration Management, Secure Software Development Management and Verification Management are practically not being addressed in any substantial measure. It is evident from the above that the ACG in its present form and alignment has a predominant 'security of infrastructure' bias rather than a desirable bias towards 'information assurance'. The Army should take serious note of this. Same is the case with DIARA at HQ IDS.

Requirement ,

The primary reason for lack of an enterprise perspective on information assurance has been the inability to view information from the strategic viewpoint and recognising the mission critical nature of this resource which is essential to success in future conflicts. Our capability to meet all information assurance objectives continues to remain fragmented because of our inability to centralise control over such information assurance assets and provide requisite collaboration between various stakeholders such as vendors/agencies undertaking development of information systems, project/programme management offices involved in deployment of information systems and users exploiting these information systems. [5]US Joint Lessons Learned Program (JLLP) highlights this necessity by observing, 'As a mission critical resource, information must be treated like any other asset essential to the survival and success of the forces. The complexity and criticality

---

[5]Joint Lessons Learned Program, Chairman of The Joint Chiefs of Staff Instruction J-7, CJCSI3150.25E, US Military, 20 April 2012, http://www.dtic.mil/doctrine/doctrine/jllpb/cjcsi3150_25.pdf

of information assurance and its governance demands that it be elevated to the highest organisational levels.'

Concerted efforts are required to address this gap in capability in order to meet all information assurance objectives. As a first step, an overall enterprise-level Information Security and Assurance Strategy (ISAS) must be defined quickly. Based on this strategy, the second step of establishing an enterprise-level Information Security and Assurance Programme (ISAP) should be taken up. The third step, an important one, is to agglomerate existing organisations like the ACG and other envisaged assets to create an Army Information Assurance Agency (AIAA) under the aegis of the DGIS to aid in implementing the above. The Army simply has to be ruthless in following such an approach, disregarding protests of loss of turf by others. If we hesitate from taking such a step, the pace of modernisation can hardly be accelerated in the realm of information warfare. Further, stunted growth will imply inadequate cybersecurity and cyber warfare capabilities, thus, severely restricting our combat potential. As part of restructuring of Integrated HQ of MoD (Army), the Army would do well to elevate the Director General of Information Systems to a Principal Staff Officer (PSO), bringing him directly under the Vice Chief of Army Staff.

The Army must align with the essential requirement of viewing information from the strategic viewpoint and recognise it as a mission critical resource. Unless this vital step is taken, shedding the baggage of legacy thinking, it will not be possible to usher in the required critical advantage in information warfare. Only then the three critical steps of defining an enterprise-level ISAS, establishing an enterprise-level ISAP and agglomerating existing organisations like the ACG and other envisaged assets to create an AIAA under the aegis of the DGIS would be taken up speedily. Similarly, at the tri-Services level, while DIARA has been in existence for some years now, the three critical steps of defining an enterprise-level Information Security and Assurance Strategy, establishing an enterprise-level ISAP and agglomerating existing organisations of the three Services need focus.

Information Security and Assurance Programme

[6]An ISAP must be developed and tailored to specific organisational mission, goals and objectives of the military. An effective ISAP constitutes a set of key elements and these individual programme elements must be integrated through common activities of establishing effective governance structure and policy, demonstrating management support to information security and assurance and integrating the elements into a comprehensive information security and assurance programme. The key elements that make up an ISAP are: Security Planning; Capital Planning; Awareness and Training; Information Security and Assurance Governance; Information System Development Life Cycle; Information Security Products and Services Acquisition; Risk Management; Security Certification, Accreditation and Assessment; Configuration Management; Incident Response; Contingency Planning and Performance Measures and Monitoring. Brief description of these key elements is as under:

- **Security Planning:** Information security and assurance planning must begin at the enterprise-level, filtering down to system-level. It is imperative to create an organisational infrastructure that supports security planning and positions the appropriate staff into key roles. This should include the chief information officer (CIO), ISAP manager or chief information security officer (CISO), Information System owner responsible for development of the information system, General Staff officer responsible to the commander of the formation for which the information system is

---

[6]Pironti, John P, 'Key Elements of an Information Security Program', Information Systems Audit and Control Association, 2005, http://www.isaca.org/Journal/Past-Issues/2005/Volume-1/Documents/jpdf051-Key-Elements-of-an-is-program.pdf

deployed, Information owner who provides or controls this information and the Information System Security officer responsible to the Information System owner for information security and assurance aspects.

- **Capital Planning:** A formal enterprise-level Capital Planning and Investment Control (CPIC) process designed to facilitate and control the expenditure of funds for the ISAP is required. An increased competition for limited budgets and resources will necessitate that available funds be allocated towards highest priority information security and assurance investments. This practice needs to be enforced at the highest level to integrate information security and assurance activities and the CPIC process.

- **Awareness and Training:** Awareness and training programmes are critical to ISAP. They disseminate security information that the workforce, including managers, require to do their jobs. These programmes will ensure that personnel at all levels of the organisation understand their information security responsibilities to properly use and protect information resources entrusted to them. Awareness is a blended solution of activities that promote information security and assurance, establishes governance principles and accountability and informs the workforce of latest news on the subject. Training imbues relevant and needed information security and assurance knowledge and skills within the workforce, supports competency development and helps personnel understand and learn how to perform their information security and assurance roles. Education integrates security skills and competencies of various functional specialties into a common body of knowledge and facilitates a multidisciplinary study of concepts, issues and principles.

- **Information Security and Assurance Governance:** The purpose of Information Security and Assurance Governance (ISAP) is to ensure that agencies proactively implement appropriate information security controls to support their mission in a cost-effective manner while managing evolving information security risks. Information security and assurance governance has its own set of requirements, challenges, activities and types of possible structures. It also has a defining role in identifying key information security roles and responsibilities, and influences information security policy development, oversight and ongoing monitoring activities. In essence, Information Security and Assurance Governance is extremely important because it is all-pervasive and provides strategic direction to the ISAP. It ensures achievement of objectives, appropriate management of risks, responsible use of organisational resources and monitoring of the efficacy of the ISAP.

- **Information System Development Life Cycle:** The Information System Development Life Cycle (SDLC) is the overall process of developing, implementing and retiring information systems. Various SDLC methodologies have been developed to guide the processes involved, and some methods work better than others for specific types of projects. Typically, the phases of initiation, acquisition, development, implementation, maintenance and disposal are addressed in the SDLC.

- **Information Security Products and Services Acquisition:** Information Security Products and Services are essential elements of an organisation's ISAP. This merits deliberation and the use of risk management processes when selecting information security and assurance products. Acquisition of both products and services bears considerable risks that must be identified and mitigated. The importance of systematically managing the process for acquisition of information security and assurance services cannot be underestimated because of the potential impact associated with those risks. Risk management processes must be utilised. Information security and assurance decision-makers must consider costs involved, security requirements and impact decisions on organisational mission, operations, strategic functions, personnel and service provider arrangements. The process of selecting information security products and services involves numerous people throughout an organisation. Each person involved in the process, whether on an individual or group-level, should understand the importance of security in the organisation's information infrastructure and the security impacts of their decisions. The personnel relevant to information security and assurance needs would include the chief information officer, contracting officer or Competent Financial Authority (CFA), contracting officer's or CFA's technical representative, IT Investment Review Board (IRB) or its equivalent, ISAP manager, Information System security officer and Information System programme/project manager (owner of Data)/acquisition initiator.

- **Risk Management:** The principal goal of an organisation's risk management process is to protect the organisation and its ability to perform its mission, not just its information assets. Risk management can be viewed as an aggregation of three processes of risk assessment, mitigation and evaluation and assessment.

- **Security Certification, Accreditation and Assessment:** Certification and Accreditation (C&A) are important activities that support risk management process, each being an integral part of ISAP. The C&A process is designed to ensure that an information system will operate with appropriate management review, and that there is an ongoing monitoring of security controls including regular re-accreditation. Security controls are the management, operational and technical safeguards or countermeasures prescribed for an information system to protect the confidentiality, integrity and availability of the system and its information. Security certification is a comprehensive assessment of the management, operational and technical security controls in an information system, made in support of security accreditation to determine the extent to which controls are implemented correctly, operating as intended and producing the desired outcome with respect to meeting security requirements for the system. Security accreditation is the official management decision to authorise operation of an information system and to explicitly accept the risk to operations, assets or individuals based on implementation of an agreed upon set of security controls. Security assessments consist of distinct tasks of assessments conducted on an information system or a group of interconnected information systems, and completing an organisation wide programme-level questionnaire.

- **Configuration Management (CM):** It ensures adequate consideration of potential security impacts due to specific changes to an information system or its surrounding environment. CM should minimise the effects of changes or differences in configurations on an information system or network. The CM process reduces risks that any changes made to a system (like insertions/installations, deletions/uninstalling or modifications) result in a compromise to system or data confidentiality, integrity or availability.

An organisation must ensure that management is aware of the proposed changes and verify that a thorough review and approval process is in place.

- **Incident Response:** A well-defined incident response capability helps an organisation to detect incidents quickly, minimise loss and destruction, identify weaknesses and restore information systems operations rapidly. Comprehensive procedures for detecting, reporting and responding to security incidents must be developed and implemented. Organisation wide capability to provide help to users after a system security incident occurs, and sharing of information concerning common vulnerabilities and threats also need to be developed.

- **Contingency Planning:** Information system contingency planning is one element of a larger contingency and continuity of operations planning programme that encompasses IT, business processes, risk management, financial management, crisis communications, safety and security of personnel and property and continuity of governance.

- **Performance Measures and Monitoring:** Performance measures are a key feedback mechanism for an effective ISAP. Key Performance Indicators (KPIs) need to be developed as part of a Quantitative Performance Management (QPM) process to monitor the ISAP.

Army Information Assurance Agency

For implementation of an organisation wide ISAP, the AIAA must have necessary enablers to provide core competencies for gestating and sustaining the ISAP. The AIAA should form the backbone of 'information assurance' in the Army through provisioning essential components for planning, preparing, executing, implementing and monitoring an ISAP. Organisational framework for the AIAA should consist of an Information Assurance Planning & Execution Division (IAPED), Army Cyber Group (ACG), an Army Information System Testing and Audit Establishment (AIST&AE) and an Army Information System Awareness & Training Establishment (AISA&TE). The AIAA could be headed by a two-star level officer under the Director General Information Systems (DGIS) who should be formally nominated as the CIO of the Army. A brief description of the components of the proposed AIAA is as under:

- **Information Assurance Planning & Execution Division (IAPED):** This sub-component should be headed by a one-star officer responsible for planning and execution aspects of the ISAP. Day to day progression of issues pertaining to security planning, capital planning, information security and assurance governance and performance monitoring should be handled by the IAP&E division.

- **Army Cyber Group (ACG):** From its existing limited role of vulnerability management, the ACG should also be responsible for the configuration management and transform to a more proactive approach in advising on information security components and overlays for information systems, reviewing of their implementation, certifying the same and also monitoring their status.

- **Army Information System Testing and Audit Establishment (AIST&AE):** The AIST&AE should form the backbone of independent validation and verification efforts to ensure information assurance control objectives of secure system software development and verification management are met. This should be achieved through gestation of Testing Competency Centres (TCC), external TCC for Non-Mission Critical Information Systems and internal TCC for Mission Critical Information Systems.

- **Army Information System Awareness & Training Establishment (AISA&TE):** The AISA&TE should be the 'people enabler' within the AIAA catering to myriad skill sets and competency generation with respect to critical areas of governance, review and compliance. This should be achieved through systematic education and skill sets generation to specific needs of sustaining information systems in the Army from inception till phasing out. This should be done through the establishment of a Governance Training Centre of Excellence and a Security Training Centre of Excellence.

The winner in future conflicts will be one who has information advantage. The military must recognise the strategic advantage that they can accrue from information resources and accord it due importance. Information assurance is vital for facing the future challenges and without information assurance, information dominance will remain a distant dream. The military's need of the hour is to get its act together quickly. The DGIS should be empowered to play the pivotal role in making sure required information assurance capabilities are provisioned and deployed in a manner that ensures desired information edge to the Army. This will be a vital step in capacity building. Challenges of the twenty-first century and increasing warfighting capabilities of our adversaries mandate an impetus for providing in-house enablers to conceptualise and gestate state-of-the-art information systems. The proposed AIAA shall be the 'enabling arm' that shall play an instrumental role in ensuring that the required information assurance capabilities are provisioned to oversee that information systems are deployed with the desired quality, at optimal cost and within the desired time frames.

10

Data Handling and Storage

The problem of storing enormous amounts of information is a global phenomenon. [1]As the militaries around the world progress more in networking, data storage is becoming increasingly important. There is a need for the Indian military to pay close attention to the details of storing data and images gathered in enormous quantities by sophisticated technologies. The requirement is not only to store the data but also to ensure that it is accessible in required time frames and of value, which may be instant availability coinciding with acquiring the information. Militaries have determined how to store the information they have, and so continue to collect more and ensure that as storage technologies mature this data can be moved to newer devices. At the tri-Services level, an assessment is needed to define how stored data can be optimised to serve everyone's needs. While this assessment is required to ensure a holistic approach, it is

---

[1]Lawlor, Maryann, 'Military Aims to Cache in on Stored Data', *Signal Magazine*, February 2001, http://www.afcea.org/signal/subjectindex/software.html

not easy as designating the metrics is a complex process. Data storage is a complicated subject and it needs to be viewed in the backdrop of issues like interoperability, information assurance and a single standard that links or supercedes several de facto standards.

[2]The Indian IT infrastructure market comprising server, storage and networking equipment will reach US $2.3 bn in 2014. [3]The market size of data centres itself is worth US $983 mn. In this era of globalisation, Information Age and e-commerce, most commercial ventures and government organisations like private companies, banks, airlines, railways etc. have put all the data in one place so that it can be used by all concerned. In India, data centres are run by companies like Tata, Wipro, Tulip Communications, Sun Microtek etc. [4]Then, the concept of cloud computing, which dates to the 1960s, has been so popular in recent times because of the economic pay-offs. According to estimates, by 2012, 80 per cent of Fortune 1000 companies were paying for some cloud computing service and 30 per cent paying for cloud computing infrastructure. Also by 2012, India-centric IT companies were estimated to be representing 20 per cent of the leading cloud aggregators in the market. Cloud computing represents a fundamental shift in computing, providing a platform for agile and cost-effective business applications and IT infrastructure. [5]Simply put, it is the convergence of virtualisation and utility-based billing. The industry logically seeks a cloud solution that supports private

---

[2]'Gartner Says India IT Infrastructure Spending Will Reach $2.3 Billion By 2014', *Gartner*, http://www.gartner.com/newsroom/id/2481916

[3]Reddy, Sridhar S, 'Emerging Trends in Data Center Industry: A Ctrl S Perspective', *Ctrl S*, http://www.ctrls.in/downloads/emerging_data_center_trends.pdf

[4]'The history of cloud computing', *China Daily*, 21 January 2013, http://www.chinadaily.com.cn/cndy/2013-01/21/content_16146386.htm

[5]Huth, Alexa and Cebula, James, 'The Basics of Cloud Computing', *US-CERT*, http://www.us-cert.gov/sites/default/files/publications/CloudComputingHuthCebula.pdf

networking, enterprise grade virtual appliances with full visibility into the life cycle of IT applications running in the cloud, adequate security and accountability. Enterprise cloud services which were sought include at a minimum—virtual networking, virtual compute, virtual storage, virtual disaster recovery and the like.

[6]A major concern in cloud computing always is that of security, although encryption is available. The buzzword here is 'trust.' It is but natural that the users are concerned about the security, access and privacy of their own data. Computer and network security is fundamentally about three goals/objectives of confidentiality, integrity and availability. Can cloud services monitor the communication and data stored between the user and the host company? Will users need to adopt community or hybrid deployment modes which are typically more expensive and may offer restricted benefits? What about acceptable levels of availability and performance of applications hosted in the cloud. Cloud providers, however, argue that because cloud computing is built on top of virtualisation, if there are security issues with virtualisation, then there will also be security issues with cloud computing. They say, even if you're not using the cloud, you already rely on and trust network service providers, hardware vendors, software vendors, service providers, data sources, etc. The fact is that while there are both advantages and challenges in cloud computing, key security issues like trust, multi-tenancy, encryption and compliance need to be examined and addressed. There are many benefits that explain why one should migrate to this new technology, and these are cost savings, power savings, green savings and increased agility in software deployment. However, cloud security issues may actually drive and define how we adopt and

---

[6]Hamlen, Kevin; Kantarcioglu, Murat; Khan, Latifur and Thuraisingham, Bhavani, 'Security Issues for Cloud Computing', *International Journal of Information Security and Privacy*, 4(2), 39-51, April-June 2010 39.

deploy cloud computing solutions. How much more security does a public cloud have compared to a community cloud and a public cloud? Can strong security controls give leeway to adopt the most cost-effective approach, assuming that massive public clouds may be more cost-effective than large community clouds which in turn may be more cost-effective than small private clouds?

Data Classification and Handling

Data Classification and Handling requires a policy for classifying and handling data based on its level of sensitivity, value and criticality to the military. Such a policy should also determine baseline security controls for the protection of data. It is generally found that printed documents in the military are replete with security classifications and detailed instructions.

Military Data Storage

[7]In a platform-centric world, since the platform is the ultimate endpiece, all systems were developed around the platform. Applications that were new were developed that most suited the application. Software was created for each platform, but they could not interchange. The centre of the world was the application and the platform and not the data that was stored in it. Conversely, NCW dictates that information is the centre of the universe. The basic capabilities that you want to develop should support this idea. The way we use data and combine it has dramatically changed the way we do things today.

Data storage is an issue which generally gets low priority in militaries despite it being vital to NCW. This could be because it gets viewed in a somewhat indirect manner. Generally, the focus is to first get the force networked and then think about other issues like data storage backup, which is a mistake. The complexity of data

---

[7] Alberts, David S and Hayes, Richard E, *'Power to the Edge, Command and Control Research Program'*, Center for Advanced Concepts and Technology (ACT), Department of Defense, USA, 2005.

storage is such that if it is not addressed simultaneously to networking the force, then major problems will invariably accrue, which in turn will affect operational efficiency. Data storage is one component of NCW, that is still maturing and involves many facets of both technology and doctrine. Military leaders of most nations recognise that in the Information Age, information is a powerful weapon. But once all the intelligence has been gathered, the reports have been filed and the collaboration has taken place, the bits and bytes must be saved in a secure place. Therefore, data storage must be viewed as important in the information assurance arsenal as well.

Going by the examples of the civilian sector, e-commerce requires that personnel in various locations have access to the data regardless of their location. These circumstances are similar to what is required by the military as it moves toward becoming a networked force. As the Services, which have been platform-centric in the past, move towards working in an information-centric environment, data storage will become a larger concern. Backup, redundancy and round-the-clock services need to be ensured. The issue is larger than just storing data. The real discriminator in currently available data storage products is the software that allows easy access as well as manages and protects the information.

Information technology has five plus two components. We have applications, processors, networks, databases and storage and then we have what people don't talk about—the people and information element. There is a need for strategic technology and data storage is one of the cornerstones. The network component is the same too. If we standardise the way we store data, the interim step is handled by tactical applications and processors, and those should be unburdened of data logistics problems.

[8]Today there are many companies that specialise in data storage

---

[8]Bailey, Alvin L, '*The Implications of Network Centric Warfare*', US Army War College Research Project, Pennsylvania, USA, 2004.

technologies and software that facilitate information sharing. They are taking steps and assessing the powerful effect an information-centric environment can have on achieving results, both commercial and military. [9]EMC, a US firm is offering a suite of products called Enterprise Storage and Enterprise Storage Networks that offer warfighting information from all sources to be stored together, managed consistently and securely made available throughout the command structure from a single technological platform. The US Department of Defence is adopting this concept. Several essential requirements for the Global Information Grid and NCW differentiate Enterprise Storage from conventional server-centric storage. The approach supports enterprise connectivity, which means that users can concurrently connect to and store and retrieve data from all major computing platforms. As the information throughout the enterprise is consolidated into one central location, military units are working in an information-centric environment so that they can leverage the data to achieve operational and mission goals rather than spend time and energy managing technology assets. In NCW, if the military has an enterprise information storage capability, it can reduce the number of required personnel to handle the data. Also, if the data is in a central location, it can be made available to more people in the right format and at the right time. So, the implications in joint operations are pretty substantial. Quick access to information could reduce the duration of an operation, saving both time and money.

The economics of data storage is also changing as devices become more affordable and the software to access information improves. The military can consider employing technologies it did not focus on in the past. More and more products are becoming

---

[9]'Accelerate your transformation to the software-defined data center with the world's most powerful, trusted, and smart family of storage products', EMC, http://www.emc.com/storage/symmetrix-vmax/symmetrix-vmax.htm

available with firms recognising financial opportunities in these technologies, which today is estimated to be a US $40 bn a year industry. Technological developments are faster because clients are searching for new ways to store and share information, wanting more and more modern, open and networked-based solution. In the future, monolithic mainframe like architecture for storage will not be able to keep pace with the net.

However, concerns about interoperability and information assurance continue to be raised by militaries. Although discussions about NCW have been taking place for several years, the focus has been on how to best network the forces, the doctrinal implications in command and control and the advantage networked communications bring to a mission. While the data storage issue has not received much attention, the interoperability of any technology adopted is an issue that must be addressed. The importance of data storage in the area of NCW is interoperability, because if you are not interoperable then you are not in the net. And if you are not in the net, then you are not in the Information Age. Archiving information is not a new problem, however, what is new is the pace of change in storage media and techniques, and that aggravates the problem. Another issue that must be addressed is security of data. There is a problem here because if you are trying to protect your information, you frequently protect that information in transit but we are moving in a time when people could have access to information you have collected. There is need to create some sort of cryptography that protects the information but allows a commander to get access to it immediately.

[10]The US military, which has invested in satellites, planes, helicopters and unmanned platforms; pursued an incredible range

---

[10]Custer, John M, 'Small devices and big data', *Armed Forces Journal*, October 2012, http://www.armedforcesjournal.com/2012/10/11458170/

of hi-technology sensors and used programmes such as Gorgon Stare, Blue Devil 2 and ARGUS, was faced with hyper-scale data of one petabyte a day to order but the best their networks could use was about 10 terabytes an hour. Therefore, the measures adopted include: 'Integrate analytics into data storage'. This technique harvests the value from raw data where it is initially stored, and thus burdens the network only with brief reports rather than hundreds of terabytes of information. Even more unique attempts at Big Data Analytics incorporate massively parallel, shared-nothing approaches that do analytics in a totally decentralised architecture. The other, 'Automated Tiering of Data Storage' works on the principle that all data is not equal, and that we can store data intelligently based upon use. Data viewed in the past ten days can be stored for instant access but data that has not been viewed in ten days can move to a lower-tier. Data that has not been viewed in a hundred days can be moved to an even lower-tier and so on and so forth.

Indian Military

Tri-Services interoperability in Indian military is yet to take off. Each service is currently moving towards networking more information. However, as HQ IDS is working towards addressing the importance of ensuring that the information can be shared between the Services, the data storage element must not be ignored and certainly requires more attention. If we are to transform from a platform-centric force to a network-centric one, then commanders at all levels must have access to real time information, and that is where the stored data becomes relevant. In so doing, the military must learn from the civilian sector and optimise upon the technologies that are already available, customising them to military requirements.

No data classification and handling policy has been defined in the military, as well as individual Services. Such a policy is required urgently with increased automation and shift to NCW. Such a policy must include baseline security controls for the protection of data.

Within the Army, data handling and data storage is being handled by the Signals instead of the DGIS though these issues are in the domain of the latter. These issues need to be made the responsibility of the DGIS and the data centres should be manned by all arms personnel. Data collated and 'filtered' laterally and vertically from designated information centre in a HQ would require being available to others in real time. Analysis and identification of how much data or information can or will flow up, down and laterally across echelons of command and organisation of boundaries is essential. An associated issue which requires attention is the security classification of data; it differs from printed material and has yet to be defined. Absence of a clear policy has resulted in data centres mushrooming all over. Computerised Inventory Control Project (CICP) under DG Ordnance Services by itself is developing a ₹400 crore data centre. Key concerns of optimal utilisation of resources, application integration, security and scalability can be met by establishing a centralised data centre for the entire Army. A Disaster Recovery (DR) Module, also to serve as an alternate centralised data centre would be required at another geographical location. The Army would need to take a call on whether to go in for a single 'Enterprise Data Network' and whether centralised data centres at Command HQ-level are also required. Data centres would need to be underground with NBC protection in Faraday modules with Electro Magnetic Pulse (EMP) protection. The Army HQ Computer Centre (AHCC) should be converted to Army HQ Data Centre (AHDC) while the CICP data centre should be converted into AHDC, integrating all requirements.

Requirement

In the US, the Floyd D Spence National Defense Authorization Act for fiscal year 2001 required the secretary of defense to prepare and deliver a report that described where all the NCW programmes were going in the grand scheme of national defence. This report was to include efforts by the Department of Defense to define NCW and a list of requirements to ensure that decisions being made today will have the Services all going down the same path. In its oversight role, the Congress proactively sought assurance that some measure of interoperability and compatibility was applied as the Services adopted new systems, including data storage devices. India lacks that type of direction from the hierarchy—both political and at MoD-level. In fact, issues of tri-Services interoperability and jointness are not being addressed in any worthwhile measure. The government needs to appoint an appropriate body under a dynamic head to direct and monitor this revolution in military transformation that is urgently required.

HQ IDS in conjunction with the three Services must address the complex issue of data storage, work out a policy, rationalise data centres and address vital issues of interoperability, information assurance including security of information in transit, real time access to information by commanders at all levels and the like. HQ IDS and all the three Services HQ must also look into the requirements of classification and handling of data. A policy in this regard needs to be urgently defined and its implementation monitored.

11

Optimising Space

Globally, the race for exploiting space-based information assets has been ongoing for past few decades. It was axiomatic that this would happen as the first satellites went up, or perhaps the plans for this had been worked out even before and were one of the catalysts for sending satellites to space. It is also obvious that countries like the US, Russia and China would be far more advanced on this count with the advantage of a headstart and having many satellites in orbit including extensive low earth satellite networks. [1]To explore the possibility of using outer space for military objectives would have been at the centre of focus for the government and military leadership of these countries. And the focus would have increased as these programmes developed. The purpose of space exploration in the US, Russia, China and other countries like France, Japan, UK, India, Israel, Ukraine, Iran, North Korea and South Korea has always been to use orbital hardware in order to boost their military potential and

---

[1]'Modern information technologies to simulate use of space-based assets', http://law-journals-books.vlex.com/vid/technologies-simulate-space-assets-56653967

defence capability. The development and employment of space-based systems throughout the 1960s and 1980s provide a compelling proof that orbital hardware was instrumental in solving the problems of global information support and communications. Such systems have no alternative today for their functional capabilities, as has been demonstrated in the post 1990s era.

As the race for weaponisation of space rages clandestinely, it must be remembered that space has numerous peaceful uses. [2]These range from terrestrial resources and environmental mapping, navigation (on land, sea and air), communications, weather forecasting to early warning of natural disasters. More than 5,000 satellites have been launched into space, of which about 10 per cent are presently functional, and a large number gets launched annually. Satellites provided a host of communications including radio, TV, internet as well as navigation facilities like Global Positioning System.

Space Based Communication Backbone

[3]Networking of multiple satellites is an efficient way of sharing communications and lowering its cost, and providing better interoperability and data integration. However, without modulations the cost of such communications can be expensive and security breaches can be expected. Then there is also the critical limitation on the raw data sent to the ground with current space communications architecture. Therefore, the requirement is to introduce powerful space-borne processing and compression technology for raw data, which can alleviate the need for expensive and expansive downlinks. Such measures would also more easily distribute processed data directly from space sensors to the end users. A space-based information network backbone can act as the transport network for mission

---

[2]Chandele, AKS, Lt Gen, Editorial, *GEO Intelligence*, Vol 3, Issue 3, May-Jun 2013, NOIDA, India.

[3]Chan, Serena, 'Architectures for a Space-Based Information Network with shared on-orbit Processing', Massachusetts Institute of Technology, Massachusetts, USA, 2005.

satellites as well as enable the concept of decoupled, shared and perhaps, distributed space-borne processing for space-based assets. Such a network backbone is designed to meet a number of mission requirements by optimising constellation topologies under different traffic models. With high network capacity availability, space-borne processing can be accessible by any mission satellite attached to the network. For creating a cost-effective network capable of supporting high data rates, the technology enabler is optical crosslinks that have been exploited to a great extent.

[4]For enhancing space-borne processing capabilities, most countries are using COTS processors that are tolerant of radiation and can be replenished every two to three years. The search for new innovations to optimise space-based information network is speeding up in order to revolutionise satellite communications and space missions. Applications already fielded include distributed computing in space, interoperable space communications, multiplatform distributed satellite communications, coherent distributed space sensing, Multi Sensor Data Fusion (MSDF) and restoration of disconnected global terrestrial networks post disaster. Consolidation of multiple communications assets into a horizontally integrated space-based network infrastructure calls for a generic space-based network backbone to be designed. A coherent infrastructure can satisfy the goals of interoperability, flexibility, scalability and it allows the system to be evolutionary. This transformational vision of a generic space-based information network allows for growth to accommodate both military and civilian demands and also lowers the price.

Battleground Space

The US, Russia and China already have ASAT weapons. The US and Russia lead in the development of military space systems, with

---

[4] Villalpando, Carlos; Rennels, David and Some, Raphael, 'Reliable Multicore Processors for NASA Space Missions', http://trs-new.jpl.nasa.gov/dspace/bitstream/2014/41793/3/11-0119F.pdf

the US accounting for nearly 90 per cent of world military space spending. [5]Militaries all over the world rely on satellites for command and control communication, monitoring, early warning and navigation. Peaceful use of outer space also include military use such as using satellites to identify targets, direct bombing raids, control and direct drone strikes to orchestrate a prompt global strike capability, anti-access strategy or ballistic missile defence. Of the some 1000 satellites reportedly operational, some 500 are in Low Earth Orbits (LEO), forty to fifty in Medium Earth Orbits (MEO) and the rest are in Geostationary Earth Orbits (GEO). The US plans to deploy the Space Based Infrared System Low (SBIRS Low) for Theatre Missile Defence (TMD). The GEO satellites also serve for communications, relays, earth observation, search and rescue, weather and research. More importantly, their early warning features track and detect ballistic missile launching. [6]At the most extreme, space weaponisation would include the deployment of a full range of space weapons, including satellite-based systems for Ballistic Missile Defence, space-based ASATs and Space-To-Earth Weapons (STEW).

[7]Interestingly, with the US becoming increasingly dependent upon its strategic dominance in space for its global military effectiveness, concerns are being raised. It is felt that there is a possibility of other nations to disrupt or degrade key space systems of the US without having to match their capabilities. Every war that the US has been involved in over the past decade and a half has been won with minimum US casualties in only a few weeks, and its success has increasingly depended on the use of thousands of precision

---

[5]Nagal, BS, Lt Gen, 'Space: The New Battleground', *GEO Intelligence*, Vol 3, Issue 3, May-Jun 2013, NOIDA, India.

[6]Sehgal, Arun, Brig, 'Militarisation of Space: India Centric Thoughts and Perspectives', *GEO Intelligence*, Vol 3, Issue 3, May-Jun 2013, NOIDA, India.

[7]Sieff, Martin, 'US Space Assets Vital but Vulnerable', 18 May 2005, http://www.spacewar.com/news/milspace-05x.html

guided weapons aided by space-based or relayed communications and tracking technology. In the 2003 Iraq War, of the 19,948 munitions used by the US, 68 per cent were laser-or GPS-guided while 32 per cent were unguided.

Though the US remains far ahead of any other country in broadband communications and surveillance capabilities, experts have warned that space-based military assets in LEO are highly vulnerable to attacks from the ground. It is easy to calculate the course of satellites because they are in fixed orbit and fire missiles or other weapons to intercept them; because LEO is around 180 miles up in space, it remains well within the range of any nation that can field intermediate range missiles. Satellites can carry the fuel and engines to manoeuver in orbit to avoid attack but manoeuvering a satellite is difficult and expensive. According to experts, anyone who can put a satellite up can have anti-satellite capability.

China

China is maximising exploitation of space at a fast pace. [8]China is reported to have established national island surveillance and monitoring system, and has completed airborne remote sensing surveillance of its 4,406 islands. According to Xinhua News Agency, the national system is mainly built on aerial surveillance with satellites, unmanned planes and cruiser as auxiliary instruments. China has an extensive satellite network and demonstrated ASAT capability in 2007.

[9]Achieving information dominance is one of the key goals for the PLA at the strategic and campaign-level. PLA recognises the importance of controlling space-based information assets as a

[8] *GEO Intelligence*, Vol 3, Issue 3, May-Jun 2013, pp 6-7.

[9] 'Chinese Military Modernisation and Force Development: A Western Perspective', Centre for Strategic & International Studies, USA, http://csis.org/publication/chinese-military-modernization-and-force-development-0

means of achieving true information dominance, calling it 'the new strategic high ground'. The PLA has developed a highly sophisticated Integrated Networks and Electronic Warfare (INEW) doctrine to organise and structure its forces for gaining information superiority. The strategic imperative for the PLA to operate in the electromagnetic domain is driving the formulation of new approach to Information Warfare (IW) termed as 'Information Confrontation' that applies 'System of Systems' operations theory to IW, viewing it as a macro system comprising discreet capabilities linked together under a single command structure and fully integrated into the overall campaign plan.

The adoption of INEW strategy as modified by 'information confrontation' suggests that the PLA is developing or has developed specific roles for Computer Network Operations (CNO) during a conflict and possibly in peacetime. Chinese strategists and operational leaders view IW as a pre-emptive tool at par with any other element of combat arms. The main role of IW is to create windows of opportunity for other forces to operate without detection or with a reduced risk of counter-attack by exploiting periods of 'blindness', 'deafness' or 'paralysis' created by information attacks. PLA leaders are conscious of the high return on investment that network paralysis warfare offers when applied to key nodes of enemy networks to achieve immense operational effect with negligible investment.

Threats and Space Systems Protection

[10]Space systems can be negated through means of deception, denial, disruption, degradation and destruction by electronic, explosive, kinetic or directed energy weapons. Capabilities to attack ground stations and communications links are increasingly available to a broad range of actors. Direct attacks on satellites require sophisticated

---

[10]'Space Security: Fact Sheet', Space Security Organization, http://www.spacesecurity.org/SpaceSecurityFactSheet.pdf

capabilities which are not widely available. Enabling technologies for space-based negation capabilities are being pursued by several states in the context of civil and military programmes. The US leads in the development of space situational awareness, an essential tool to support space negation. The ability to detect, withstand and recover from attacks against Earth-based or space-based segments of a space system are crucial to their protection. USA and Russia lead in general capabilities to detect rocket launches, while the US leads in the development of advanced technologies to detect direct attacks on satellites. While protection of satellite communications links is poor, improving and protection of vulnerable satellite ground stations remains a concern. Protection of satellites against some direct threats is improving, largely through radiation hardening, system redundancy, and greater use of higher orbits. Following a direct attack on satellites, Russia and the US are leading in the development of capabilities to rapidly rebuild space systems.

Satellite Communications

Satellite Communications (SATCOM) are increasingly becoming important in fast moving operations not only for Non-Line-of-Sight (NLOS) communications but also where large and varied forces, including Strike Corps and their various battle groups, are engaged in manoeuvres and intense battle. This helps to optimise NCW capabilities to multiply the effect by converging at a time and place of one's choosing over long distances. Advantages of SATCOM are being exploited by terrorists too, as is apparent from the large number of Inmarsat and Thuraya radio sets captured in anti-terrorist operations in India. With flexible operational services and compact ground terminals, SATCOM services offer attractive solutions for military users. Dedicated military satellite networks are augmented by commercial services like Demand Assigned Multiple Access (DAMA) controlled networks that offer the user total control of the space link.

SATCOM provides terrain independent communications, flexible networking and direct link to final destination without reliance on radio relays.

With Apple, INSAT satellite series, ASC Network and a whole gamut of active spacecraft in GEO, India has come a long way since 1981 when ISRO first experimented with geosynchronous telecommunications relays. Indian GEO platforms perform the dual functions of communications and earth observation. India's first exclusive military satellite, GSAT-7, was successfully launched by European space consortium Arianespace's Ariane 5 rocket from Kourou spaceport in French Guiana in August 2013. It has a footprint of around 1,000 nautical miles over the Indian Ocean. The requirement was first projected by the Navy a few years back. Though the Army and the Air Force woke up late, they will partially share the bandwidth of INSAT-7 till dedicated Army and Air Force satellites are launched.

Globally, communication satellites are functioning on C, Ku, Ka, S and L bands. C Band is widely used and proven, has a large bandwidth and no rain attenuation. Ku Band has similar advantages plus Communication on the Move (COTM). However, in both C and Ku Band, the equipment is not easily portable and prone to mechanical failure due to size, weight and movement. The sub-assemblies are large and so the installation time is considerable. In the Ka Band, broadband communication is possible, power consumption is low, use of solar panels is possible and feasibility of COTM exists. However, this system gets adversely affected by snow, rain and heavy clouds. Space segment in Ka Band is presently limited in India. The S Band is meant only for radar communications and hence, not discussed further. L Band is far more reliable especially under NLOS conditions as accurate pointing is not required between satellite and user terminals. The system is highly portable and man-portable option is available. It is lightweight with no moving parts, quick to deploy with rapid connectivity, not affected by weather conditions, has low power requirements, lower attenuation and better range. However, L Band has less bandwidth availability compared to C, Ku, Ka bands and has interference with terrestrial communications. L Band services include Standard IP, flexible bandwidth based on usage, user control and spot beams.

A dispassionate analysis of the various types of satellites would indicate that L Band is most suited for non-terrestrial communications in the Tactical Battle Area (TBA). L Band terminals are truly portable, plug and play, reliable on account of integrated design, communications for land, sea and air applications, flexible and have low first-time investments. For some inexplicable reason none of the indigenous satellites are L Band, even though the initial investments compared to other satellites is less. The dedicated military satellite (INSAT-7) to be launched too, will not be L Band. Though a number of foreign satellites on L Band have footprints over India, use of a foreign satellite for operational military communications raises legitimate apprehensions of security. Foreign firms argue that security while using a foreign satellite can be made foolproof if the security gateway is positioned within Indian territory and indigenous security solutions developed by CAIR are superimposed. However, this requires a thorough analysis. Under our existing regulatory regime, any company even wishing to provide subscription-based television services to customers in India must uplink from Indian territory and must use INSAT or a satellite system approved by Department of Space (DoS). Approval for using a foreign satellite by a DTH service provider is given rarely for short term (with ISRO approval) until INSAT is able to launch an additional capacity.

Role of Satellite Communications

Satellite communications play a vital role in hugely varied terrain (mountains, deserts, jungles, urban areas) and areas with poor or non-existent telecom infrastructure/terrestrial communications, especially with increased requirements of COTM. Military applications of satellite communications include collaborative battlefield planning using interactive whiteboards and video conferencing, live video feedbacks from UAVs, reconnaissance units/combat camera teams of troops in contact/Special Forces missions etc. 'Any Voice to Any Voice', push to talk to landline phone and merging of voice, data and video on a single network is facilitated. Modern satellite communications cater for increased mobility, are easier to use and with increased throughput are usable at lower levels for VTC, situational awareness and sensor information.

Indian Army Scene

Although a VSAT network is in the process of being established as part of project Air Defence Control & Reporting System (ADC&RS), the Army needs to shed its inhibition of treating SATCOM as 'an expensive and rare commodity' with continued reliance on terrestrial communications for vital systems like the Battlefield Surveillance System (BSS). Not only communications have to be foolproof, future requirement of a capable NCW capable force needs to be well understood. While there is merit in relying on terrestrial communications, satellite communications must be scaled selectively at lower levels as well after thorough appreciation of future battlefield requirements. Stryker Units of the US are already provided satellite terminals down to platoon-level. Problems with current communication systems are poor spectrum utilisation, low data transfer rates, predefined mission planning and intentional and unintentional jamming environment resulting in poor communication performance. Modern satellite communications help overcome these problems. The rule should be 'communications without break' using terrestrial communications 'where feasible and when not adversely affecting network centricity'.

The Army needs to plan on selective SATCOM both in TBA and upward connectivity including redundancy in critical situations in war and national security. The TCS is to cater for TBA requirements primarily for offensive operations by battle groups of both the Strike and Pivot Corps, albeit its induction and fielding is inordinately delayed. Fielding of information systems like the BSS, Battlefield Management System (BMS), ADC&RS, Artillery Command Control and Communication System (ACCCS) and their integration to the TacC^{3}I system through the CIDSS, coupled with the requirement of foolproof communications in fast moving intense battle environment necessitates due prominence to SATCOM. This is also pertinent to F-INSAS and particularly Special Forces detachments operating deep inside enemy territory for where compact lightweight portable SATCOM terminals would be essential.

What the Army Needs

As part of static communications, the Army requires linking of remote areas through point to point bridges and other networks. Then it requires connectivity to Indian contingents on various UN missions. Satellite overlays are needed for static communication network and logistic nodes plus peacetime establishments. Then comes the requirement for information/intelligence gathering and dissemination. For mobile communication in the TBA, the requirement is: Transportable satellite ADC&R and strike formations; satellite overlays for TCS and Out of Area Contingency and Disaster Management. Requirement of ground segment of satellite communications are: Static and transportable satellite terminals, which will provide voice, data and video conferencing facilities to the users; satellite based WAN on static VSATs to meet the networking requirements of logistics and training units; man-portable satellite terminals with voice and data facilities and networks with secrecy devices. For the space segment of satellite communication, it is prudent to have transponders on multiple civilian satellites in C, S, K and Ku Band and they should have flexibility and redundancy.

IRNSS

As part of Indian Regional Navigational Satellite System, ISRO was to get the IRNSS fully operational by 2012. However, of the seven satellites that were to be launched, the first was launched in 2013. IRNSS will have a coverage of 1,500 kilometres and beyond with 10 metre accuracy over India and 20 metre accuracy over the Indian Ocean Region.

Counter Space

Offensive counter space involves the use of lethal or non-lethal means and is conducted to achieve five major purposes of deception, disruption, denial, degradation and destruction of space assets or capabilities. Conversely, defensive counter space encompasses: Reducing and precluding the effectiveness of an adversary's counter space operations; preserve own ability to use space systems; active and passive defence; capability to steer satellites to safety; and catering for adequate redundancies till hardening/deterrence capability is achieved. We must be able to fully exploit space technologies, which are: In communications, steerable beams to provide communication in less frequented areas. High bandwidth, polar micro communication satellite network, cryptography, data compression and satellite crosslinks; in surveillance, RS satellites with adequate coverage and revisit capability with selectable multi-imaging techniques; in targeting, integrated precision positioning of targeting and telemetry signals; in early warning, sensors for Inter-Continental Ballistic Missile (ICBM), Terminal Ballistic Missile (TBM) and satellite launches; spy satellites for high resolution imagery high sensitivity RF, IR, laser receivers for ELINT; sub-metre resolution regional navigation system for navigation and positioning; finally, application embedded software for automation of functions like detection, cyberattacks, cartographic application etc.

Future Course of ISRO

As part of remote sensing satellites, ISRO is looking at sub-metre multi-spectral imaging capabilities, imaging-on-demand facilities, high agility and intelligent reconfigurability. In small and scientific satellites what is being looked at is lightweight, low-cost platforms, multi-spacecraft mission capability, constellation reconfigurability, single transponder microsats for internet trunking, COTS components/materials usage and launch on demand. In communication satellites: Constellations of small satellites against/complementing large platforms; larger and powerful spacecraft with lower cost per transponder/full use of frequency spectrum at the available orbital location; variety of services in a single satellite—typically two 3-tonne satellites combined as one 5-tonne satellite resulting in 25 per cent cost reduction; larger EIRP, better SFD, ability to provide spectrum on demand; and longer life. Technologies for launch vehicles planned to be developed are re-entry materials, thermal protection systems, high thrust cryogenic engines, air breathing engines for re-entry vehicles, semi cryogenic engines, ion propulsion systems and nuclear propulsion systems.

Required Focus

It would be prudent for the government/MoD to consider the following:

- Removing the asymmetry vis-à-vis China in exploitation of space. Chinese space assets are as vulnerable as that of any country. India must also develop ASAT capability and establish a space-based communication backbone optimising processors and compression technologies.
- Development and launch of L Band satellites by ISRO for military SATCOM due to advantages, including COTM, as discussed above. If agreed upon, the ground network and terminals will need to be developed concurrently.
- Due to high costs of development and operations of satellite

systems, outsourcing of services is considered by many nations. There is a need to examine implications of hiring a foreign L Band satellite for military communications (till all the requirements of the Army are met through ISRO satellites). Such an exercise should include feasibility of ensuring foolproof security with the security gateway positioned within India and with superimposed security solutions developed by CAIR. Obviously, such an arrangement would require firm commitment regarding availability of space segment at all times and under all conditions, aside from security issues being suitably addressed. Need for such examination is necessary in order to leapfrog into netcentricity, which is presently too distant.

Additionally, the Army needs to focus on the following issues:

- Review the communication philosophy, appropriately incorporating SATCOM to meet requirements of netcentricity in future wars.

- Requirements of SATCOM must be identified, link data rates determined, design and size of each link decided and the rationale documented. Analysis should take into account Orbit, RF Spectrum, Data Rate, Duty Factor, Link Availability, Link Access Time, Threat etc.

- System chosen should be flexible and modular, catering to data security through encryption, spatial, time and satellite diversity, frequency hopping and interleaving.

- Terminals chosen should be flexible and modular; easy to move between networks, with easily adaptable throughput and easily switchable from military to commercial satellite, should need arise for the latter. Time required for training and operations should be minimal. They should need limited logistic support, easy for undertaking field repairs and with swappable modules.

- Compression of data is essential since digital image files in particular are large and at times there will be need for sending more data than what the bandwidth accommodates; bandwidth is limited by the link equation. Therefore, the best compression technologies must be exploited.
- Spectrum management is very important. Experience shows that at any given time, large bandwidth of the entire RF frequency (up to 100 GHz) remains unutilised. System chosen should be able to capitalise on unutilised bandwidth.
- SATCOM systems, other than in L Band, are bulkier and hence, the Army has been contemplating use of Wimax in TBA. This must be given adequate thought especially from security point of view. Because of doubts of foolproof security of Wimax, foreign armies are using Wimax only for Logistics Nets and *not* for Operational Nets.

During times of war and conflict, only the best communication system will succeed. SATCOM is vital for an NCW capable Army. The satellite can support thousands of terminals in net-centric system formation. For redundancy, the network can use multiple satellites or alternate communications. Military grade on-the-move SATCOM ground terminal must provide continuous connectivity in conditions where commercial terminals usually fail. They must be able to automatically and rapidly recover from signal blockages, due to man-made objects, terrain/foliage, weather and other atmospheric effects. Such terminals are designed to work with a wide range of military and commercial satellite services, including wideband commercial LEO, MEO and military GEO satellites, as well as Protected (Advanced EHF) Narrowband Satellite communications. The Army should review its communication philosophy accordingly and cater for requisite SATCOM with alacrity, boosting its quest of acquiring true NCW capabilities.

12

Modelling and Simulation

Militaries around the world use virtual, live and constructive simulators for training their personnel. These ensure soldier safety and cost savings in terms of preventing wear and tear of original equipment, fuel and man-hours. Simulators are also used to enhance doctrinal, tactical and strategic warfare skills of military personnel. Virtual simulators (used indoor) modelled on the original equipment are designed for training under varied computer-generated virtual 3-D environments. Live simulators, used for outdoor training, have laser-emitting devices and sensors to gauge performance of the trainee. [1]Constructive simulators are meant for computerised war-gaming simulators, using logical and mathematical modelling to represent the dynamics of combat training.

[2]Modelling and simulation is about getting information on how something will behave without actually testing it in real life.

---

[1]Sishtla, Srinivas, 'Indian Military Training and Simulation Market: All Set to Experience an Exponential Growth', Institute of Defence Studies and Analysis, New Delhi, India, 25 August 2009, http://www.frost.com/sublib/display-market-insight.do?id=178013878

[2]Maria, Anu, 'Introduction to Modeling and Simulation', USA, 1997, http://www.inf.utfsm. cl/~hallende/download/Simul-2-2002/Introduction_to_Modeling_and_Simulation.pdf

It involves using models including emulators, prototypes, simulators and stimulators, either statically or over time, to develop data as a basis for making managerial or technical decisions. [3]Modelling is a process of producing a model which is a representation of the construction and working of a system of interest, the purpose being to enable analysts predict effect of changes to the system. A good model is a judicious trade-off between realism and simplicity. Simulation practitioners recommend increasing the complexity of a model iteratively. An important issue in modelling is model validity. Model validation techniques include simulating the model under known input conditions and comparing model output with system output. Generally, a model intended for a simulation study is a mathematical model developed with the help of simulation software. Mathematical model classifications include deterministic (input and output variables are fixed values) or stochastic (at least one of the input or output variables is probabilistic); static (time is not taken into account) or dynamic (time-varying interactions among variables are taken into account). Typically, simulation models are stochastic and dynamic.

Simulation of a system is the operation of a model of the system. The model can be reconfigured and experimented with; usually, this is either impossible, too expensive or impractical to do in the system it represents. The operation of the model can be studied and hence, properties concerning the behaviour of the actual system or its subsystem can be inferred. In its broadest sense, simulation is a tool to evaluate the performance of a system, existing or proposed, under different configurations of interest and over long periods of real time. Simulation is used before an existing system is altered or a new system built, to reduce the chances of failure to meet specifications, to eliminate unforeseen bottlenecks, prevent under or over utilisation of resources and optimise system performance.

---

[3]ibid

The terms modelling and simulation are often used interchangeably. Use of modelling and simulation by militaries has been more in the domain of individual weapon platforms; training through simulated firing to cut costs. However, in advanced defence organisations and militaries like the US, modelling and simulation is part and parcel of the acquisition procurement strategy. Modelling and simulation is optimally used to conduct events and experiments that influence requirements and training for military systems. As such, modelling and simulation is considered an integral part of systems engineering of military systems. [4]Other application domains, including NCW, are progressing rapidly.

[5]The military simulation and virtual training market in 2011 was assessed at US $8.75 billon and estimated to grow in multiple sectors over the coming years. Military simulation products and equipment are sought by defence ministries on a global scale in order to provide lifelike training for service men and women in the land, air and sea domains. As a consequence, military simulation is no longer confined only to the field of aircraft simulation, although the aviation sector will continue to play a major role within the industry and in the wider market, simulations linked to combat operations, medical treatment, maintenance, modelling and unmanned systems, mean that virtual training can now take place through multiple scenarios and on multiple platforms. In addition, the development of transportable simulation systems and the introduction of day/ night, urban and rural and adverse weather condition technology, means that training can be tailored to specific needs and individual missions.

---

[4]'Empowering the Army', Army Modeling and Simulation Office, US Army, http://www. ms.army.mil/

[5]'The Military Simulation and Virtual Training Market 2011-2021', *Visiongain*, August 2011, http://www.visiongain.com/Report/654/The-Military-Simulation-and-Virtual-Training-Market-2011-2021

[6]Globally, there are 111 associations, organisations and committees, forty centres and groups, and twenty-seven military organisations dealing with modelling and simulation. In India, we have the Centre for Mathematical Modelling and Computer Simulation (C-MMACS) and the Indian Simulation and Gaming Association (INDSAGA) but these have not been capitalised upon by the military in any way. Military organisations for modelling and simulation exist in the European Union, NATO, Australia, Canada, Italy, Republic of Korea, Sweden, Turkey, the UK and the US. USA takes the cake with some thirteen modelling and simulation organisations spread over the Department of Defense, Army, Air Force and Navy-marines.

Modelling and Simulation for NCW

Significant technological advances over the past few decades have fuelled the continual and rapid development of an information-based world. NCW has become the buzzword and is becoming a popularly shared vision and rallying cry for force transformation of militaries around the world. An essential element in fully implementing this network-centric way of thinking is to develop useful measures to help gauge the effectiveness and efficiency of both our military networks and strategic NCW doctrine.

To establish metrics for measuring NCW is the foremost vital step for modelling and simulation in this field. In terms of establishing metrics for the three domains spanning NCW, the physical domain is the easiest since it covers the traditional environments of land, sea, air and space in which conflict typically occurs, and is home to the platforms and communications networks of a given military force. However, the physical domain provides an incomplete picture in capturing the complex interactions and outcomes of real warfare.

---

[6] Modeling and Simulation (M&S): Associations/Organizations/Committees, http://www.site.uottawa.ca/~oren/links-MS-AG.htm

This is the primary reason for including the information and cognitive domains in the conceptual framework of NCW.

The information domain, in which information is created and manipulated also encompasses all the means of conveying the decisions, plans and orders that translate a cognitive response into physical actions. Hence, the cognitive domain is the locus of the functions of perceiving, making sense of a situation, assessing alternatives and deciding on a course of action. Though this domain exists within the mind of the warfighter with all the attendant intangibles, metrics for this domain are by far the most difficult to assess. However, methods and tools for analysing human decisions are being constantly developed and improved upon. These, coupled with evaluation of artificial intelligence, would eventually forecast leadership decisions made in conflict situations. This would be the eventual conquest over the cognitive domain.

The task of determining appropriate and measurable metrics for NCW is highly complex and there are no set rules. Infrastructure Performance, Battlespace Awareness, Battlespace Knowledge, Exploiting Battlespace Knowledge and Military Utility are general categories under which fall the more precisely defined metrics for NCW. However, there have been suggestions to describe metrics for the characteristic 'speed of command', force agility and the ability to amass effects, the 'degree of autonomy' aspect of self-synchronisation, the level of shared situational awareness, the conduct of effects-based operations, reach back operations, information superiority, the degree of interoperability and mutual trust. Others propose that the key characteristic of network centricity is the broadening the focus of warfighter away from the individual, unit or platform concerns to give primacy to the mission and responsibilities of the team, task group or coalition. Some have proposed more quantifiable metrics for measuring NCW in the form of connectivity, reach, richness and characteristic tempos. Whatever the case, quantifying this 'broadening of focus' is a difficult problem, especially when one tries to do so in a sense that is independent of a specific scenario.

US Military

[7]To understand the mechanics of modelling and simulation for NCW, we may look at a model in the US. The US Navy leads the US military joint programme Network Warfare Simulation (NETWARS). In this capacity, the Navy has been developing communications models of Navy systems for use within the NETWARS programme, integrating federate communications infrastructures developed in NETWARS and other simulation systems to leverage the strengths of each simulation system and support analysis of NCW concepts in operational scenarios. The Navy Simulation System (NSS) and NETWARS are fully integrated. NETWARS is a discrete event simulator developed using the Optimized Network Engineering Tool (OPNET) Development Kit (ODK). It has been designed to analyse military communications networks through the use of reusable communications device models (CDM), military doctrine and network traffic information in the joint arena. NETWARS consists of four functional elements: Database libraries; Scenario Builder; Simulation Domain and Analytical Tools. The simulator uses database libraries such as Communications Device Model (CDM); Operations Facilities (OPFAC); Organisation and Information Exchange Requirements (IER) to obtain detailed information about the communications systems used during the analysis. The CDM library contains the models that have been developed by the Services to represent the protocols and functionality that is found in real physical devices.

---

[7]Alspaugh, Chris; Davé, Nikhil; Hepner, Tom; Leidy, Andy and Stell, Mark, 'Modeling and Simulation in Support of Network Centric Warfare Analysis', Command and Control Research and Technology Symposium (CCRTS), San Diego, USA, 2004, http://www.dtic.mil/cgi-bin/GetTRDoc?AD=ADA466029

The Scenario builder is used to define how the OPFACs, Organisations, links and IERs are to be used during the simulation. OPFACs and Organisations can be developed and links can be assigned. Mobility can be given to Organisations to represent the real time movement of units throughout the course of the simulation. IERS are associated with devices and the time in which the IERS are sent is also defined here. Periods of failure and recovery of OPFACs are also specified within the Scenario Builder. The Simulation Domain provides for the conversion of scenario information into a common Scenario Definition File (SDF). In theory this can be submitted to any simulation engine that supports the SDF format. At this time, the SDF file is fed into the OPNET simulation engine. This is a commercial COTS discrete event simulator that is used to assess the traffic flowing across the network. The Analysis Tools provide a way to examine the results of the simulation. They allow the analyst to examine the Measures Of Performance (MOPs) that were collected during the simulation. A sampling of the MOPs that NETWARS monitors during a simulation include link utilisation, throughput, message delay, message completion and failure rate. The programme can also collect node-level statistics or other custom statistics if the user incorporates the collection of these statistics within the appropriate CDMs. NETWARS, also, can make use of a High Level Architecture (HLA) interface to permit communications between the NETWARS simulator and other simulators.

The NCW analysis branch of the Space and Naval Warfare (SPAWAR) Systems Centre has supported simulation-based assessments for over ten years. SSC has an extensive library of communications models to support analysis that range from capacity planning, to prototype modeling, to military communications planning and doctrine development. These models are interoperable and reusable, and can be leveraged to support. They include the Link-16 model suite, which is a tactical message exchange system.

These models represent communications characteristics of the Tactical Data Information Link (TADIL) J communication system.

[8]The US Air Force is using Agent Based Model (ABM) and System Effectiveness Analysis Simulation (SEAS). In ABM, complex, real-world systems are modelled as collection of autonomous decision-making entities, called agents. Each agent individually assesses its situation and makes decisions based upon its own set of rules. Agents may execute various behaviours appropriate for the system they represent such as sensing, manoeuvering and engaging. SEAS is a constructive, agent-interaction based simulation designed specifically for exploratory analysis of transformational, information-driven warfare across surface, air and space domains. SEAS has the ability to model the presence and interaction of a large variety of unique agents within a combat mission scenario. It is built around three simple entities: Agents, devices and environments. Essentially, agents interact through the use of devices (weapons, sensors, communication) with each other and the environment. Conflict outcomes emerge from these resulting interactions. Agents are logical members acting within the combat mission scenario. They can be units, such as a brigade or multi-ship formation of planes, or subunit members such as a vehicle, individual plane or satellite. Devices are entities such as communications devices, sensors and weapons. The environment is the battlespace, which consists of events, locations, terrain, weather, jamming and day/night characteristics.

UK Air Force

The modelling and simulation team at Royal Air Force (RAF), UK are using a twenty-first century crystal ball to help predict the future state of the RAF's Typhoon fleet. It encompasses aircraft maintenance, technical support, training and asset management to support

---

[8]Honabarger, Jason B, BS, 'Modeling Network Centric Warfare (NCW) with the System Effectiveness Analysis Simulation (SEAS)', Air Force Institute of Technology, Ohio, USA, March 2006, http://www.dtic.mil/cgi-bin/GetTRDoc?AD=ADA446395

operations of Typhoon aircraft. Each service element interacts, influences and is, in turn, influenced by the others. The modelling and simulation team have deployed an Integrated Typhoon Availability Service Performance Simulation (TPS) toolset. It is basically a computer game where scenarios are fed into the simulation software and it predicts an outcome. The software is very much like a football managing game, you pick the players, the formation and the style of football and then you get an outcome. If any of the elements are changed, the output is different. It helps make quantifiable decisions, has ability to boost service and improve operational readiness.

Indian Military

As the Indian military modernises and inducts new platforms and complex surveillance and delivery means, the need to transform training philosophy, modify and improve the present training infrastructure and methods, introduce new training technologies to save costs and get better results in terms of skill sets and efficiency has never been greater. [9]The Indian Army says, 'Simulation advances will transform military planning and training. Today, virtual reality simulations can enable soldiers to train in high fidelity mock-ups, at substantial reduction in risk and spending. There is a need to exploit a range of tools and products that will enhance the Army's capabilities in the domains of training, development, acquisition and decision support.' [10]Growth of simulators has been catalysed because of induction of sophisticated equipment with mechanised forces in the lead followed by all other fighting and supporting arms through simulators like Drona, Simulated Fire or SimFire, driving simulators etc.

---

[9]'Technology Requirements of the Indian Army', http://www.ciidefence.com/pdf/Future_Technology_Requirements_of_the%20Indian_Army.pdf

[10]Sharma, Abhimanyu, 'Military Training & Simulation: Part II', *South Asia defence & Strategic Review*, 12 August 2012, http://www.defstrat.com/exec/frmArticleDetails.aspx?DID=375

In 1991, the Army established the Simulator Development Division (SDD), which designed the Drona simulator and took up other projects like a Gypsy simulator and a even a BMP II simulator despite lacking production facilities. The second and more challenging is the fact, that as the Army begins to rely more on simulators, the level of realism grows exponentially, requiring domain specialists to design and develop simulators based on existing software platforms. This a challenge that also confronts the DRDO as it aims to develop simulators at the Institute of Systems Studies and Analysis. But capabilities of ISSA itself are poor considering the way it handled ASTROIDS as discussed in Chapter 5. It first fielded a model that went defunct because of numerous shortfalls, and then failing to upgrade its software even after a decade, forced the project to be foreclosed. Similar has been the case with the Infantry tactical training software developed by ISSA for the Army's Wargaming Development Centre (WARDEC). It was found to be so rudimentary that the Army perforce is now looking at the private sector. Unlike the Navy and Air Force, the Army is yet to commence using simulators. The concept that has gained popularity abroad calls on third-party vendors to establish, maintain, upgrade and in certain cases, even staff the simulation centres where troops come and train on the latest equipment, without burdening the Army's budgets. If such a system were to be introduced in India, it would help the Army conserve vital resources as it also provides an avenue for gainful employment of retired Army officers.

Simulators have enabled the Indian Navy to practice the challenges of operating at high seas through simulated training in navigation, watch-standing, ship-handling, deck equipment operation, engine management, firefighting and weapon systems management. The Navy also has introduced simulators that account for vagaries like vessel characteristics, wind, tide and swell, passage of other vessels, seabed shape etc. Ship handling simulators enable the crew to hone their skills as well as prepare for mission specific challenges by practicing them beforehand. Along with training costs, one of the biggest advantages of naval simulators is their ability to reduce crew wear and tear. With the need to practice live drills reduced substantially, the crew now spends lesser time training at sea, giving them more time ashore. The Navy has also begun the process of digitising its training content to provide an immersive learning environment.

By allowing the Air Force to practice flying sophisticated aircraft in an accurate representation of the real world, simulators fill what is potentially a vital training deficiency. Simulator training on the use of aircraft systems including handling weapons, radars, jammers, improves a pilot's proficiency and enhances overall operational effectiveness of assets. A simulator flight generates a large number of data points that can be analysed either by computer algorithms or by the instructors themselves, who now need not divide their time between flying and instruction. Feedback is instantaneous and can be reviewed along with the pupil during the debriefing. With the simulator's ability to recreate specific situations, trainers can get straight to instructional part of a flight without having to go through the rigours of startup, taxi, circuit and so on. Also, simulators enable the crew to identify and remedy potentially life-threatening situations and address problems like spins, stalls and severe turbulence.

Joint Operations

With the military eventually adopting an NCW doctrine of operations, they will have to train their officers to perform command and control functions in the face of an information overload. Though this has spurred the demand for constructive simulators, the segment is still nascent and characterised by low contract values. The low contract value makes the segment unattractive for most large players and leaves the door wide open for indigenous Small and Medium Enterprises (SMEs). However, we have to get over the mental block of such projects not being hijacked by the DRDO and Public Sector Undertakings (PSUs) whose track record and accountability is poor, to say the least. Ironically, while simulators have proved to be invaluable to the IAF's flight training process, their lack of interoperability has hindered their use to train officers for joint warfare, a crucial component of NCW. With limited indigenous capability and a lack of global standards, limited interoperability is a small price to pay for the mental peace of having a simulator that is supported by the Original Equipment Manufacturers (OEMs) as long as the equipment remains in service.

In the present context, no worthwhile dedicated effort in modelling and simulation is underway within the military or for that matter indigenously, though we have had seminars and exhibitions in the national capital periodically. From 21-22 June 2012, [11]India hosted a conference-cum-exhibition on Military Simulation & Training. [12]This was followed by a DEF SIM 2012, a Military Modeling and Simulation Solutions seminar on 23 August 2012 in New Delhi hosted by EDS Technologies, Bengaluru and VT MAK. But in most seminars the policymakers and those who are to execute them generally attend the inaugural or concluding sessions of the seminar. One example is the mapping and Geospatial Intelligence mess that India is in (discussed in Chapters 6 to 8) despite international-level GEOINT seminars year after year. Coming back to modelling and simulation for the military, the Army does not even have COTS tools to simulate behaviour of mobile networks in differing terrain, weather and electronic warfare environment. Simulation and modelling requires development of algorithms

---

[11]Military Simulation & Training India on 21-22 June 2012, Air Force Auditorium, Subroto Park, New Delhi, India, http://www.shephardmedia.com/static/files/download/MSTI-Brochure-2012.pdf

[12]DEF SIM 2012, Military Modeling and Simulation Solutions Seminar, New Delhi, 23 August 2012, http://www.edstechnologies.com/Mailer/aug12/news_3.html

which is a specialised task. Prediction models are even more difficult to develop. These projects need to be outsourced and unless we do so, we are unlikely to improve the state.

Systems of Interest

[13]There is no doubt that most international products in modelling and simulation are of US origin, some of which are being used by the UK, Australia and Taiwan. India should, therefore, optimise on defence cooperation and strategic partnerships with these countries in this domain. A system of interest at the tactical-level could be the JANUS, which has been used by NATO and its allies and the Joint Conflict and Tactical System (JCATS). At the operational-level, systems of interest, could be the Joint Theatre Level Simulation (JTLS), a campaign-level training system also acquired by Pakistan and the Extended Air Defence Simulation (EADSIM). At the strategic-level, systems of interest could be the Joint Warfare Simulation (JWARS) jointly developed by the US and Rand Strategic Assessment System (RSAS).

Requirement

The government, MoD and the military need to look at the requirement of modelling and simulation in holistic manner and pursue a focused road map including nominating segment of the private sector to leapfrog such technologies, providing them directions, finances and infrastructure with timelines to acquire such capability. Considering the threats that India faces, that call for a synergised national response, it is prudent to establish a National War Gaming Centre, preceded by a tri-Service Military Wargaming Centre to exploit the NCW capabilities being developed. The need for urgency was never more considering we are way behind time. The following merit focused consideration in this regard:

---

[13]Anand, Vinod, War-games, Modeling, Simulation, *SP's Military Yearbook*, SP Guide Publications, New Delhi, India, 2008-2009.

- The military must understand the potential of the systems that can be fielded, shortcomings in the existing systems (there are many) and scope of synergy in current activities. All this requires a group of professionals specialising in these aspects in a sustained manner and having had experience in this field.

- MoD and the military will need to jointly define a policy for Modelling and simulation encompassing objectives, organisations, responsibilities, resource allocations etc. A common policy framework would provide the necessary momentum in enhancing the military's modelling and simulation capabilities.

- HQ IDS will need to exert in development of the systems and essential interoperability between the three Services.

- Then would come the requirement to spread awareness and organise joint training from tactical-level upwards. The subject should be a part of training curriculum of military's training institutions.

- The military should also establish links with civil institutions to refine operational research. The yet-to-be-established Indian National Defence University (INDU) should play a major role in this.

13

Human Resources

[1]The importance of 'man behind the machine' is as relevant to NCW as before. [2]Discussion of NCW challenges military assumptions about the individual and collective employment of people in delivering effects. The success of NCW rests on the idea that information is only useful if it enables more effective action. But the key to success of NCW is not technology but unambiguously the people who will use it—the human dimension. The human dimension is based on professional mastery and mission command, and requires high standards of training, education, doctrine, organisation and leadership. The dimension is about the way people collaborate to share their awareness of the situation, so that they can fight more effectively. It requires trust between warfighters across

---

[1]Human Factors in Network Warfare Symposium, Australian Government, Department of Defence, Defence Science and Technology Organization, Sydney, 1-3 May 2006, http://www.dsto.defence.gov.au/events/4659/

[2]Fogarty, Gerard, Brigadier, 'Progressing the Human Dimension of NCW in the ADF', Canberra, 2006, http://www.dsto.defence.gov.au/attachments/Keynote%20Address%20Brigadier%20Fogarty_Progressing%20the%20Human%20Dimension%20of%20NCW%20in%20the%20ADF.pdf

different levels and trust between warfighters and their supporting agencies. The human dimension of NCW is complex and difficult to conceptualise. Most defence forces around the world are struggling with the issue and have achieved breakthroughs in varied measure.

[3]One of the tenets of NCW is that increased information-sharing enhances not only the quality of information but also encourages collaboration and increases shared awareness. The power of NCW lies in a robustly networked force and its ability to move information, manage it and thus, perform missions more effectively. Information and more importantly knowledge, lies at the heart of NCW. While the role of information in NCW is clear, there remains much to understand about how human beings share, absorb and make sense of the available information and subsequently make decisions based on that information. For example, an increase in the amount of information available to commanders does not necessarily result in improved knowledge nor help them make better decisions. The premise that more information is better is not always true. While gathering information enhances intelligence, it also must aid understanding and make sense. Coupled with the aspect of information are issues like understanding the power of the applications itself, one example being the Decision Support System properties and limitations. [4]The purpose of an information management strategy or plan is to improve participants' ability to find the data they need and to understand that data when they receive it. Data and information must be visible, accessible, understandable, trusted, interoperable and made available in response to user needs. Moreover, the individuals or units must be able to obtain all the

[3]Ali, Irena, 'Is NCW Information Sharing a Double Edged Sword—Voices from the Battlespace', 2006, http://www.dsto.defence.gov.au/attachments/Ali_Is%20NCW%20 Information%20Sharing%20a%20Double%20Edged%20Sword%20-%20Voices%20 from%20the%20Battlespace.pdf

[4]Renner, Scott, 'Net-Centric Information Management', 2005, http://www.dodccrp.org/ events/2005/10th/CD/papers/348.pdf

data/information needed and retrieve that information repeatedly for verification.

[5]In network-centric environments, team decision-making and situation assessment are distributed in both time and space. Shared understanding among team members with regard to the impact, importance and quality of relevant information items (eg. sensor outputs, text documents, images, message traffic, web pages) is a critical element in the selection of an effective course of action. Here, focus required is—what is the minimum information that needs to be exchanged for shared understanding to occur; how do we capture that information and how should it best be displayed? Distributed teams that communicate asynchronously require a knowledge management plug-in tool that will convert, encapsulate and tag a group member's subjective understanding of a complex information item into an iconic representation that represents various information parameters. Icons can be automatically generated from an abstraction template completed by a team member for each decision-relevant information item. These tags can then easily be electronically exchanged among the team, improving shared understanding, consensus building and information fusion among group members and significantly reducing the valuable decision time typically consumed by conflict resolution. [6]In the US, NATO and coalition forces, chat has become a dominant communication vehicle. Based on a compilation across multiple stakeholders and inputs, chat tools are developed to meet user needs

---

[5]Cowen, Michael B and Fleming, Robert A, 'Collaboration in Command and Control Environments: Exchanging Iconic Tags of Key Information', USA, 2006, http://www.dsto. defence.gov.au/attachments/Cowan%20&%20Fleming_Collaboration%20in%20 Command%20and%20Control%20Environments%20-%20Exchanging%20Iconic%20 Tags%20of%20Key%20Information.pdf

[6]Catanzaro, Jean, 'User-Centered Design Requirements for Maritime Chat Clients in Network-Centric Warfare', USA, 2006, http://www.dsto.defence.gov.au/attachments/Catanzaro%20 &%20Smillie_User-Centered%20Design%20Requirements%20for%20Maritime%20 Chat%20Clients%20in%20Network-Centric%20Warfare.pdf

in a dynamic, maritime coalition environment. Implementation of these chat tools enable effectiveness and efficiency because they facilitate situation awareness; allow users to gain access to chat information in a timely manner; provide optimal record keeping (archive) storage and retrieval capabilities, facilitate timely message composition, transmission and receipt; provide user identification, presence and status information and provide system administrator the capabilities to ensure chat tools run smoothly and that updates can be implemented in a systematic and effective manner. Standard implementation of these requirements is a clear step toward efficient use of collaborative technology and interoperability between sites. [7]The intricacies of what the human dimension is required to deal with NCW environment may be gauged from the [8]US Navy System Engineering, Acquisition and PeRsonnel INTegration (SEAPRINT) programme to develop processes, architectures and policies that lay the groundwork for successful implementation of effective systems that are configurable, reusable and scalable. Implementation of SEAPRINT rests on defining and practicing the processes and establishing policy to support the systems engineering requirements for human system integration.

[9]The ability to achieve human performance improvement is

---

[7]Narkevicius, Jennifer McGovern and Owen, John E, 'Human Systems Integration Enhancing Performance in Network Centric Warfare: The SEAPRINT Perspective', USA, 2006 http://www.dsto.defence.gov.au/attachments/Narkevicius,%20Stark%20&%20Owen_Pending%20Official%20Release_Human%20Systems%20Integration%20Enhancing%20Performance%20in%20Network%20Centric%20Warfare.pdf

[8]Dolan, Nancy and Narkevicius, Jennifer McGovern, 'Systems Engineering, Acquisition and Personnel Integration (SEAPRINT): Achieving the Promise of Human Systems Integration', Washington DC, USA, 2006, http://ftp.rta.nato.int/public/PubFullText/RTO/MP/RTO-MP-HFM-124/MP-HFM-124-01.pdf

[9]Pester-DeWan, Joanne and Oonk, Heather, 'Human Performance Benefits of Standard Measures and Metrics for Network-centric Warfare', USA, 2006, http://www.dsto.defence.gov.au/attachments/Pester-DeWan,%20Oonk%20&%20Smillie_Human%20Performance%20Benefits%20of%20Standard%20Measures%20and%20Metrics%20for%20Network-Centric%20Warfare.pdf

limited by the lack of infrastructure or process to support coordination among various organisations in tracking and using a common set of human performance measures, metrics and results. The metrics ontology tool is a system that provides researchers, system developers, designers, trainers and others concerned with human performance measurement, access to commonly defined and structured measures, definitions and measurement results. The Human Systems Integration (HSI) community is collaborating to develop the tool as a way to encourage a common language of measures and measurement and to support the sharing of newly created measures and previously collected data across Navy organisations. The ontology tool allows users to: Browse measures and metrics; retrieve previous assessment results and share assessment results with other user communities. In essence, the ontology tool provides assessment communities with a shared, integrated repository of measures, metrics and assessment data for accurately and completely tracking performance measures and assessment results.

[10]NCW is currently the dominant logic of military operations—a major conceptual platform from which modern militaries will address the challenges of future operating environments. In militaries of advanced countries, road maps for NCW are progressed in two dimensions simultaneously to enhance the overall warfighting capability; one is the network dimension, referring to the physical systems providing connectivity between sensors, commanders and those involved in engaging the adversary, and the second dimension is the human dimension road map. [11]During the period

---

[10] Warne, Leoni; Bopping, Derek and Ali, Irena, 'NetworkER Centric Warfare: Outcomes of the Human Dimension of Future Warfighting Task', Australia, 2006, http://www.dsto.defence.gov.au/attachments/Warne,%20Bopping%20&%20Ali_NetworkER%20Centric%20Warfare%20-%20Outcomes%20of%20the%20Human%20Dimension%20of%20Future%20Warfighting%20Task.pdf

[11] Dekker, Anthony H, 'Revisiting 'SCUDHunt' and the Human Dimension of NCW: Some Thoughts', Canberra, Australia, 2006, http://www.dsto.defence.gov.au/attachments/Dekker_SCUDHunt.TTCP.pdf

2000-02, Computer Network Assurance Corporation, USA and ThoughtLink Inc conducted 'SCUDHunt' game-based experiments to explore Shared Situational Awareness (SSA) in an NCW environment. It proved that 'good' and 'bad' teams had a greater impact on mission effectiveness than any technology factor. It also showed that major contributors to mission effectiveness were good team dynamics, professional mastery and ability to use the technology. Of these, team dynamics was the most important. The analysis also illustrated the fact that the often lauded SSA is not an end in itself—poor teamwork and insufficient levels of professional mastery can lead to teams sharing agreement on incorrect assessments of the situation, sometimes with tragic results. The conclusions underlined the importance of the human element in the networked future force and supported emphasis required on professional mastery.

It is also important for us to understand that the information revolution and networked environment give rise to various entities. Hence, it is imperative that Services retain core competency and have the ability to integrate with other domain specialists. Therefore, there is an urgent requirement to establish concept development centres, which will require system analysts and programmers.

Indian Military

Our military has yet to fully realise the essential requirement of viewing information from the strategic viewpoint and recognise it as a mission critical resource. Had we realised this, we would have adopted a top-down approach since not only do we need a synthesis of communications and information, we would have speedily addressed alterations in our concept of operations, doctrine, organisation, force structure, psyche, leadership and associated changes in logistics, education and training will also be required, all of which need to be concurrently addressed in order to acquire, to build and enhance

NCW capabilities within the constraints of development and implementation time. This requires that while concerted efforts have to be made at all levels to conceptualise, define, induct, implement and operate the new systems in a defined period of time, human resource has to be built to handle and optimise those systems, the latter demanding professional mastery, awareness and best training, education, doctrine, organisation and leadership.

In the Indian military, majority of work in the NCW domain has been conducted with the network dimension in mind though in disparate fashion on individual Service basis. Efforts to progress the human dimension have followed the same route and are somewhat embryonic. Progress in the tri-Service network domain including defining an NCW doctrine has not happened. It should be quite clear that even if an NCW doctrine is defined, it will not be possible to implement it since HQ IDS does not wield sufficient power in absence of a CDS with full operational powers over the three Services. The unpalatable fact is that in its present shape, the Services do not care much about HQ IDS and installation of a permanent chairman of Chiefs of Staff Committee (COSC) with perfunctory powers would hardly change the equation. Also, not many would know that the organisation of HQ IDS reportedly worked out in seven days, and has hardly come up with an organisation that is administratively sound. For example, the appointment of assistant chief of Integrated Defence Staff (strategic operations) tenable by a two-star rank officer who usually assumes this appointment after command of a division, is not authorised a clerk and not even a runner, the only authorisation being a personal assistant who may just know rudimentary computers.

Apart from this, the requirement for the Indian military to harness disparate efforts and to generate some forward movement in terms of human resource development for NCW was debated in depth at HQ IDS during 2005-06. Accordingly, a case was taken up with the MoD to outsource a holistic study on the requirement of IT training for the military as a whole. This would have helped establish the human dimensions of the networked force and develop a detailed understanding of how to achieve the greatest synergy between our systems and human resource and initiate changes to optimise education, training and development, taking into account information sharing and collaboration (organisational structures and command and control). This would cover operators, managers, commanders, administrators, system analysts, programmer and so on and so forth. The MoD accorded approval within a short span of time but inter-Services differences scuttled the issue yet again, highlighting the need for a CDS who is empowered to make a single decision for the military. By the time the MoD approval arrived, the Chief of Air Staff took the stance that such a study was not required in the first place. Such was his vehemence that he would not even agree to let the study be outsourced (that would have gone to the private sector), analyse it when completed and then decide/ discuss whether the Air Force wanted its implementation or not. The end result was that despite the MoD sanction, the study was not outsourced. So much so for tri-Service networking! In due course, the Navy went ahead and raised its own IT cadre. The Army continues with its own IT training in-house and by outsourcing through private institutions.

The major problem of the military not accepting the demand for specialisation is much more in the Army than in the other two Services. Despite understanding and experiencing that operational information systems take a long time to develop, the Army's military Secretary's Branch is stuck on granting extension beyond the tenure of three years in Delhi only on 'case to case basis'. So, we have a situation where an officer will land up in Delhi because he has to be given a posting here, is placed in information systems where he learns the job and is moved out much before his full potential can be utilised. To top this, many officers are posted to mark time after having been approved for the next rank. There are numerous cases of officers of Lieutenant Colonel rank having been posted to these project management organisations and moving out even before completing one year. These 'post office' postings are not confined to young officers alone. The appointment of the DGIS itself has been treated as 'transit' with many moved out in a short span of time to other appointments. If the military would look closely at the organisation and truly treat information as a strategic resource, the appointment of DGIS should actually be made tenable by an officer who has commanded a Corps, to provide the necessary vision. In absence of such an arrangement, we may take wrong decisions, like when the BSS was conceived, it was only planned at Divisional Headquarters and above levels, not recognising that the cutting edge in the TBA operates under the Brigade HQ and requires real time/ near real time information. Even the fact that many a Brigade HQ had already improvised and was operating ad hoc, BSS was ignored. BSS at Brigade-level was added at much later date. Similarly, the test bed for the ADC&RS was conceived in isolation by the Army without integrating the Air Force, knowing full well that overall responsibility of air defence lies with the Air Force and not the Army. It is only later that the Air Force was integrated into the test bed for ADC&RS. Similarly, the requirement to make the appointment of Additional Director General of Military Survey tenable by an all arms officer who has commanded a Division has been covered in Chapter 6 earlier. Similarly, the appointment of DGIS should be made tenable by an officer who has preferably commanded a Corps in order to provide the strategic direction for development of information systems.

But it is not the project management systems of DGIS that are affected by this system of three years posting, extendable on 'case to case basis'. Most surprisingly, niche areas in Delhi like the

Army Software Development Centre (ASDC), under DGIS and Army Cyber Group (ACG) under the Signals officer-in-chief are also meted out the same treatment. Limited tenures are also the bane of Defence Information Assurance & Research Agency at HQ IDS. The least that the military could have done was that let such organisations be manned by women officers for their complete service tenure. The need for an IT cadre itself should be examined on priority because every time a new operational system is tested in field, the Corps of Signals projects demand by hundreds for increase in manpower. NCW is supposed to decrease overall manpower but that will happen only in the long run. The experience in militaries of developed countries has been that when new systems are introduced, the manpower required actually 'goes up', which is followed by a gestation period, levelling out and finally decreasing. Therefore, it would be prudent for the military, particularly the Army to examine the need for a separate IT cadre. While the military is yet to define a policy for Modelling and Simulation, for developing, establishing, fielding such systems, as well as in the field of operational research, there can be no shortcut to specialisation. All this will require a group of professionals specialising on these aspects in a sustained manner, having had experience in this field.

Next is the vital requirement for senior appointments at policy-making levels to understand technology/make credible efforts to understand technology. In no case does this imply an M Tech or B Tech degree. As part of the Chinese RMA, only those PLA officers are posted in policy-making appointments related to NCW who understand technology. In the case of our military, at least two senior officers, one Vice Chief of Army Staff and one Deputy Chief of Army Staff acknowledged that they do not understand technology, as mentioned under acknowledgements to this book. It is for similar reasons, the development of Phase 3 of F-INSAS (Computer and radio subsystems plus software integration) is continued by the Infantry, simultaneous to the development of BMS by DGIS, despite the fact that BMS caters to the entire Army at unit and regiment-level including the Infantry. The Navy's Operations Branch is headed by the Assistant Chief of Naval Staff (Operations and IW) but the Military Operations does not even have an Information Systems cell, though there are two sections dealing with Signals and manned by officers from the Corps of Signals. The tussle between applications and networks is a global phenomenon that is happening in foreign militaries as well—the chicken and egg syndrome. Both have to depend upon each other, like pillars and roof of a building. The synergy, therefore, has to come through an architect. It is for this reason that the appointment of Deputy Chief of Army Staff (Information Systems & Telecommunications) was created; to ensure concomittant development between applications and networks. In the beginning, there was no staff authorised under him to advise him on Information Systems. A start was made in this direction many years later by creating a cell under him to render advice but more needs to be done. In the absence of expertise on Information Systems in Military Operations, case files are stuck for months, shuttling between Operations and Signals. Interestingly, the case file to affix responsibility for who is to ensure security of Army intranet shuttled within Army HQ for full three years. Still decisions have not been fully ensured and hence, e-learning programmes have not taken off. Military Operations had charged the Additional Director General Electronic Warfare (ADG, EW) to look after cases of Information Systems but that has not worked well. In one case, the ADG (EW) wanted the DSS module of CIDSS scrapped because he felt that the DSS should decisively give out the course of option to be adopted. As mentioned earlier, mere M Tech and B Tech degrees are not enough. Having maximum number of such degree holders, the Signals Directorate is still not automated. These are issues that need to be looked into by the military and the Army.

The importance of the human dimension to developing an NCW capability is well known, but unfortunately it remains the least understood and researched domain of NCW. Our difficulty in conceptualising the human dimension and its complexity is arguably the reason it has fallen behind NCW related material enhancements. The fact that material enhancements alone will not generate the desired mission effects is well understood, but coordinated work to effectively link doctrine, organisation, training, material, leadership and personnel enhancements has been exceptionally slow in case of the military.

14

Hurdles to Informisation

The term 'Informisation' draws global attention not because the Chinese coined it but because of the institutionalised accelerated pace at which the PLA is getting 'informised' as part of RMA. The Pakistan Army is following suit with focused investments in relevant sectors, aided amongst other factors by the US (courtesy Global War on Terrorism) and China. In the Indian military, there is 'informisation' to some extent but the Army is lagging behind. Significantly, the planned reorganisation of Pakistan Army's General HQ envisages merger of the Communications Branch into Information Systems Branch. In contrast, the Indian military has progressed little beyond making statements viewing information from the strategic viewpoint and recognising it as a mission critical resource. In its concerted efforts to modernise, it is essential the military, particularly the Army, must align with this truth and stop treating information just as another resource. Unless this vital step is taken, shedding the baggage of legacy thinking, ushering required critical advantage in Information Warfare (IW) would not be possible. With existing mindsets, even the goal of achieving NCW capabilities faces constant slippages, thereby negating capacity building. There is an urgent need for internal reform for which some hard decisions are required to be taken.

Informisation

What constitutes informisation in a military? Essentially, it is the confluence of Operational Information Systems (OIS), Management Information Systems (MIS) and Geographical Information Systems (GIS) and their integration with systems like the Electronic Warfare System (EWS) and Electronic Intelligence (ELINT), Logistics Management System (LMS), Information Assurance including cybersecurity, automation and digitisation, e-war gaming and simulation, e-learning, e-procurement, online audit and the like. All this with matching communications that enable real time/near real time exchange of information which meet all military requirements including fast paced manoeuvres.

In the context of the Indian Army, at the heart of informisation lies the Tactical Command, Control, Communications and Information (TacC^{3}I) system. Within the TacC^{3}I, the subsystems of Command Information Decision Support System (CIDSS), Battlefield Support System (BSS), Artillery Command Control and Communications System (ACCCS), Air Defence Control & Reporting System (ADC&RS) and Battlefield Management System (BMS) are all bound by the CIDSS as the backbone, also configured to integrate systems like the EWS and ELINT. Subsystems of TacC^{3}I are in varied stages of implementation; from already fielded to Request for Proposal (RFP) for Phase 1 yet to be issued. Above the CIDSS (top end being the Corps HQ) is the Army Strategic Operational Information Dissemination System (ASTROIDS). MIS too, are in various stages of development. Communications planned for exchange of information are the Tactical Communications System (TCS) and the Defence Communications Network (DCN), other than the Army intranet on which applications like Army Wide Area

Network (AWAN) function. The TCS is also to integrate mobile phone network of the Army. War-gaming, simulation and Information Assurance are at nascent stages. E-learning has hit a roadblock with Army intranet still not made secure, especially, in terms of last mile albeit AWAN messages are being sent on it without a security solution in place. Additionally, online audit has been fully introduced in the Air Force but the IA is not yet game for it because of security concerns.

The fielding of TacC^{3}I has been facing excruciating delays with stiff resistance from certain quarters; not recognising mission criticality of information and fearing loss of individual turf and dilution of comfort zone. Though the Directorate General of Information Systems is given the charge of facilitating transformation of the IA into a dynamic network-centric force, achieving information superiority through effective management of information technology, it cannot do this in requisite time without cooperation within the Integrated HQ of the Army. Presently, DGIS lacks breathing space and hierarchical intransigence to this state of affairs, adversely impacting informisation of IA.

GIS and Military Survey

In 2009, during an international seminar in New Delhi, a participating Deputy Inspector General (DIG), Border Security Force (BSF) demonstrated a GIS that had been introduced in the BSF. He stated this happened after BSF acquired a how-to-go-about it from Military Survey (MS). It is a shame that GIS is still to be fielded by the IA. A GIS policy and Tri-Service Common Symbology was issued by the IA in 2009, though Military Survey was under Military Operations till 2005. This was after some eighteen months of comprehensive study, members included engineers and sister Services. Yet, engineers tried to block its issuance even at the last moment fearing loss of turf. It is for the same reason that the draft RFP for an Enterprise GIS (approved in principle for fielding up to Corps-level) kept circulating within Army HQ for eleven months with little accountability by those stonewalling it. The RFP was eventually issued but has not been followed up since 2009. Post fielding of Enterprise GIS, the next step is to take it down to the Tactical Battle Area followed by the establishment of a Spatial Data Infrastructure (SDI).

Military Survey came under DGIS in 2005 on express directions of the Raksha Mantri in order to ensure synergy of GIS with the OIS and MIS. Mapping Policy of India clearly states that Survey of India is responsible for issue of all maps within Indian borders including Defence Series Maps (DSMs) but provision of maps by SoI and their updating is years behind schedule. The reason quoted is lack of manpower in the SoI and delayed book debit payments by the government. This despite a subunit of Military Survey at Agra with dedicated IAF reconnaissance aircraft (they do not fly within 10 kilometre of border) undertaking SoI tasks without payments being made to MoD in return. Instead of ensuring that SoI delivers on their mandate, Military Survey has been involved in production of DSM series maps (task of SoI) and physical survey for collection of attribute data, albeit conveniently ignoring insurgency areas. Undue emphasis on physical survey needs to be viewed in context of satellite imagery and modern technology plus the fact that no physical survey is possible for transborder maps where actual battles will be fought. Computer-based digital techniques of handling geospatial terrain information and GIS as an efficient decision support tool can efficiently handle modern warfare. Technology provides new methods for preparing digital terrain data for both digital and paper maps.

Military Survey follows an old system of 'reverse deputation' with SoI. This was perhaps relevant in initial stages when expertise in survey was limited. It needs to be replaced by a simple 'deputation' of three to five years, especially considering that this reverse deputation was hardly practised in the last twenty-five years. Military Survey on an average has been holding only fifteen to twenty survey trained officers (these too, mostly trained by IA, not SoI) against an authorised strength of 103 survey trained officers over the last twenty-five years. Engineers hardly send officers on survey courses due to shortages and such training is organised 'after' the officer joins Military Survey. In the bargain, a large number of engineer officers in SoI, in cushy jobs, have attained the rank of major general—far more than what they could hope within IA. In 2009, when turnover issue was raised by the DGIS, SoI offered major general rank officers from SoI to replace colonel-level officers in Military Survey. Incidentally, officials of the Ministry of Science and Technology privately admit that SoI is one of the worst managed organisations. The merger of the survey sections at Corps HQ-level with IITs for creating GIS subunits was approved through an Army study in 2009. Yet, this approved study was not implemented, Military Survey moved out from DGIS and now a fresh study has been ordered. How this can be justified to the MoD is difficult to comprehend.

BMS vs F-INSAS

The BMS and F-INSAS programmes are under concurrent development; BMS under DGIS and F-INSAS under Infantry. BMS was conceived at battalion/regiment-level pan Army (including for the Infantry) and comprises communication, non-communication hardware and software. The system will be further integrated with the TacC^{3}I through the CIDSS. Quite logically, Phase 3 of F-INSAS (Computer subsystem, radio subsystem, software and software integration) should be part of the BMS. However, the Infantry has been given the go-ahead. A separate project of software and communication integration by Infantry will be retrograde, delay overall netcentricity pan Army, incur additional avoidable costs and defeat the very purpose for creating DGIS. Already considerable work in the fields of applications has been done by the latter in addition to completing Phase 1 of CIDSS and Battlefield Surveillance System.

Squabbling on delimitation between the BMS and F-INSAS have cost a delay to Phase 1 of BMS by more than 12 months (an inexcusable folly surprisingly supported by Military Operations). The Signals (whose role is to provide communications to Battalion HQ and above) supported the Infantry, sensing they would get a role for provisioning communication equipment for F-INSAS. If Infantry is to incorporate Situational Awareness and GIS, then it amounts to 're-inventing the wheel' and yet, another project to integrate the F-INSAS with BMS with additional expenditure and time. Logic demands that for Phase 3 of F-INSAS, Project Management Organisation (PMO) F-INSAS should be placed under DGIS as part of BMS. Later, BMS can be developed for Mechanised Infantry in dismounted role.

Cyber Security and Information Assurance

In most armies of the world, Information Assurance (including cybersecurity) is handled by Information Systems. Signals quote that Signals officer-in-chief was nominated chief security officer in 2004. The fact is that DGIS came into being only in December 2004. This anomaly needs to be rectified. Relating existing capabilities of Army Cyber Group (ACG) to the information assurance control objectives clearly shows that only issues relating to Personnel Management and Vulnerability Management are being addressed and that too, in limited form. Other information assurance objectives of Configuration Management, Secure Software Development Management and Verification Management are practically not being addressed in any substantial measure. ACG in its present form and alignment has a predominant 'security of infrastructure' bias rather than required bias towards 'information assurance'. IA should take serious note of this. The capability to meet all information assurance objectives continues to remain fragmented because of our inability to centralise control over information assurance assets and lack of requisite collaboration between various stakeholders such as vendors/agencies undertaking development of information systems, project/programme management offices involved in deployment of information systems and users exploiting these information systems.

IA needs to take concerted and early steps to address this gap in capability for meeting all information assurance objectives. An overall enterprise-level Information Security and Assurance Strategy must be defined quickly. Based on this strategy, an enterprise-level Information Security and Assurance Program (ISAP) should be taken. It is important to agglomerate existing organisations like the ACG and other envisaged assets to create an Army Information Assurance Agency under the aegis of the DGIS to implement the ISAP. Unofficially, officers of ACG themselves opine that ACG should have been part of DGIS. The IA simply has to be ruthless in following such approach, disregarding protests of loss of turf by others. If we hesitate to take such a step, the pace of modernisation can hardly be accelerated in the realm of IW. Further, stunted growth will imply inadequate cybersecurity and cyber warfare capabilities, severely restricting our combat potential. It would be prudent to make the DGIS a Principal Staff Officer, bringing him directly under the vice chief. This would also help integrate systems like EWS and ELINT controlled by Military Operations and Military Intelligence respectively. In cybersecurity training, we should graduate from existing elementary to cybersecurity war games where networks need to be kept operational under battle conditions while hackers try to infiltrate with methods that are likely to be used by our adversaries, such as flooding servers to block them, planting viruses etc. It goes without saying that Information Dominance must be an essential element of our war doctrine. We must be able to protect own information systems, attack/influence information systems of adversaries and leverage own strengths to gain decisive advantage in a battlespace where national security is threatened.

The ACG should expand to take on the role of AIAA. To ensure an organisation wide ISAP, the AIAA must have necessary enablers to provide core competencies for gestating and sustaining the ISAP with sub components of Information Assurance Planning & Execution Division (IAPED), ACG, Army Information System Testing and Audit Establishment (AIST&AE) and Army Information System Awareness & Training Establishment (AISA&TE). Presently, numerous applications are coming up throughout IA, some of them without adequate security solution and without reference and clearance from ACG. Once total networking is achieved this would jeopardise security.

Data Handling and Data Storage

Even some Signals officers are surprised that Data Handling and Data Storage are being handled by the Signals instead of the DGIS, these issues being the domain of the latter. Data collated and 'filtered' laterally and vertically from designated information centre in a HQ would require being available to others in real time. Analysis and identification of how much data or information can or will flow up, down and laterally across echelons of command and organisation of boundaries is essential. An associated issue which requires attention is secure classification of data. This differs from printed material and has yet to be defined. Absence of a clear policy has resulted in data centres mushrooming all over. By itself, Computerised Inventory Control Project (CICP) under DG Ordnance Services is going for a ₹400 crore Data Centre. Key concerns of optimal utilisation of resources, application integration, security and scalability can be met by establishing a centralised data centre in Delhi. A Disaster Recovery module to serve as an alternate centralised data centre would be required at another geographical location. Centralised data centres at Command HQ-level would be needed. Data centres should be underground with NBC protection in Faraday modules with Electro Magnetic Pulse (EMP) protection. Army HQ Computer Centre

(AHCC) converted to Army HQ Data Centre (AHDC) need to be made all arms and placed under DGIS. Eventually, the under construction CICP Data Centre should be converted into AHDC, integrating all requirements.

Communications

Army Intranet: The Army intranet is still to be made secure, as last mile security has not been ensured. This needs to be done at the earliest. Fixing responsibility with Signals took three years. There should be no ambiguity that security of communications is the responsibility of Signals Officer-in-Chief. Securing Army intranet would open floodgates of e-learning including online instruction of courses and professional exams. IA also needs to review its policy of not providing Army intranet to tri-Service training institutions like Defence Services Staff College (DSSC) and College of Defence Management (where almost 90 per cent staff and students are from the Army) and disconnecting HQ Integrated Defence Staff (IDS).

TCS: Approved by three successive RMs, TCS has still not seen the light of the day. It needs to be accelerated. The void has affected test beds for TacC^{3}I subsystems. Designating a Corps as test bed has little meaning if adequate communications cannot be provided. Till date, all test beds have been truncated with obvious disadvantages.

Communication Support for TacC^{3}I Subsystems: Lack of synergy between Signals and IS delays projects as the former, relying on existing terrestrial communications and legacy radios, needs repeated convincing with accredited loss of time. Repeated arguments delay initiation of projects. Initially, Signals were loath to accept Software Defined Radios (SDRs), albeit they were not to 'replace legacy radios in totality'; SDRs with dual wavelength are available and can also communicate with legacy radios. The main problem with Signals is that they look at bandwidth requirements based on communications in truncated test beds whereas prudence demands we cater for communications for fast-paced manoeuvre battles. It is no secret that IA woke up late to the need of a dedicated satellite. As the largest user of space, the IA should have taken the lead to project such requirement. Study on bandwidth requirement for IA has been completed only recently at the behest of DGIS.

Communication Data Network System (CDNS): The CDNS commanded by a director-level officer is under DG Signals on the plea that it was created out of former manpower of Signals. Its role includes assisting TacC³I components in selection of communications equipment, interact with TCS, advise component system on the choice of access media for connectivity to CIDSS, work on standards and protocols, advise on network security, network and spectrum management plus facilitate single window interface of DG Signals with all TacC³I components. Placement of CDNS within DGIS will actually ensure greater jointness and viewing information from the strategic viewpoint instead of a biased Signals view.

DCN: Project DCN is to go down to Corps/equivalent levels of the three Services. While IA is in charge of the project, providing the strategic communication highway, little is happening on the Services handshake for exchange of information. Presently, the process of defining common standards and protocols for the Services is progressing at a snail's pace at HQ IDS. It would also be prudent to go in for underground DCN nodes instead of over ground, as planned, since these will be lucrative targets in the event of war.

Manpower and PMOs

The failure to view information from a strategic viewpoint has led to 'post office temporary posting' in DGIS. The last three director generals (including the present one) have been in transit, two of them holding the appointment for seven to ten months. Deputy DDG-level officers heading PMOs are also being shifted out after a couple of months. Director-level officers, already approved for promotion, get posted in, only to move out within months. All this adversely affects projects that have long gestation periods. Expansion of Army Software Development Centre (ASDC) approved in principle years has not been implemented. This needs to be expedited. The ASDC should be converted to an Army Software Management Centre and made part of Army Information System Testing and Audit Establishment (AIST&AE) under the AIAA, as proposed above. IA should consider long and dedicated tenures at least in organisations like ASDC and ACSE including short service women cadres doing their entire/bulk service in such appointments.

War-gaming and Simulation

Wargaming Development Centre (WARDEC) located within DGIS enclave is under Army Training Command (ARTRAC), not even under DG Military Training. The systems and applications are vintage, using limited maps depicting terrain on the side of the border. The Wargaming Centre at Chandimandir, Panchkula has progressed little. War-gaming at Army Headquarters and at formation headquarters-level should be planned based on the TacC^{3}I systems with integrated GIS of transborder terrain where operations are envisaged. This should be the forte of the DGMT/DGIS while WARDEC could modernise for training on counter-insurgency/counterterrorism plus war-gaming in training institutions like Army War College.

Technology Voids

Electro Magnetic Pulse (EMP) attacks (nuclear and non-nuclear) at critical times are a reality. This is a serious challenge. With technological advances and globalisation trends, systems and applications will continue to have foreign content, no matter how miniscule. Adversaries can embed vulnerabilities (Trapdoors, Trojans etc) in both hardware and software. We need to build capability for checking/testing and safeguarding against these. Centre for Artificial Intelligence and

Robotics has to exponentially increase its capacity to develop new and varied algorithms in order to keep pace with rapid induction of TacC^{3}I and other systems. Scientific Advisory Group must find ways and means to accord approvals in a telescoped time frame compared to the months/years being taken presently, resulting in inordinate delays in test beds. Also, we need to speedily advance our chip manufacturing capabilities, a sphere in which we are decades behind China, and which has serious implications for network security.

Focus Required

The military needs to urgently review how it wants to treat information. Prudence demands we adopt an approach to establish information superiority at the earliest and plan to maintain/improve upon this edge. In the present environment, this appears a distant dream. There is a crying need to take an overall view and make top-down changes. This alone can usher in the required information revolution replacing existing evolutionary approach.

15

Managing Transformation

There has been much talk of transformation of the military in India, especially after the Kargil conflict. A dispassionate analysis, however, would indicate that ground-level translation of the many recommendations for transformation has largely been cosmetic or at best marginal, despite so many years having elapsed. We do not appear to be making a holistic effort to learn from the transformation of militaries in countries like the US, the UK, Germany and China. The same ambivalence relates to revamping the national security architecture, notwithstanding the recent study headed by the chairman, National Security Advisory Board (NSAB) and recommendations of which may also be only partially implemented and that too, over a prolonged period going by earlier experiences. The profound changes in the last two decades of the twentieth century should have radically altered our perceptions towards the nature of future conflict and mechanics of its resolution, including the ensuing sea change in the way nations perceive themselves and each other in the overall international system. Whatever the future, with comprehensive security getting more and more complex coupled with technological advances, the military will continue to be an important and critical element of national power. To that end, India must acknowledge that the Indian military establishment requires creative adaptation, fundamental changes and determination to be able to respond effectively to the nation.

There is a need to study the issues pertaining to management of transformation and towards realising joint capabilities in the Indian armed forces. In this context, there is a need to draw upon both the organisational and military theories being refined and practised by others. This, then, should lead us to the change management required for transformation of the Indian armed forces and thereon to a road map to affect the same. We need to examine some of the organisational theories that primarily come up in the face of international business competition but are also relevant to the military, as also catalysts and theories for military transformation. Then, there is the need to examine models of military transformation, concluding how India should manage strategic military transformation of the armed forces.

Theories

- **Organisational and Business Oriented:** Advances in information technology and accelerating foreign competition had started emerging as challenges as early as the late 1970s and the 1980s. The most affected being the business world, they started looking for ways and means to meet these challenges in order to regain and retain competitive advantage in respective businesses. As a result, a multitude of theories, strategies and techniques of management emerged throughout this period. All these theories sought to describe the appropriate steps necessary to renew and re-energise the basic business organisation. [1]John P Kotter, professor at Harvard Business School, propounded eight steps (termed Kotter's Theory)

---

[1]Kotter, John P, *Leading Change*, Harvard business Review Press, USA, 2012.

that are required to transform an organisation: One, to establish a sense of urgency, contending that failure to instill a sense of urgency is the biggest mistake leaders make when trying to transform an organisation and that establishing a sense of urgency and eliminating complacency are crucial to gaining the cooperation needed to drive the transformation process. Two, to form a powerful guiding coalition of the right people who enjoy a great deal of trust and share a common objective since transformation requires a powerful force to sustain the process and no individual, regardless of formal or informal power or weak committees, can lead or manage transformation by himself. Three, to create a vision, since a vision performs the triple tasks of providing general direction of transformation simplifying the number of decisions a business needs to make, provides motivation to the people and coordinates all actions. Four, to communicate the vision since communication is essential in ensuring that people within the organisation have a common understanding and shared sense of commitment to the future. Five, to empower others to act on the vision since effectively empowering subordinates results in the four-fold advantage of removing structural barriers, provision of required training, aligning the organisational system to the vision and in dealing with troublesome supervisors. Six, to plan and create short-term successes since that provides credibility to the transformation and helps sustain it over the duration of the process. Seven, to consolidate and produce further changes/transformation optimising on credibility afforded by short-term victories to tackle bigger problems within the plan. Eight, to institutionalise new approaches where leaders need to anchor change within the organisation's culture to ensure long-term success of the transformation effort. [2] A second example is Robert Miles, an academician and professional consultant on change and transformation, who defines four fundamental attributes (termed Miles Theory) associated with successful transformation: First, the ability to thrive on energy with direction; Second, a total system perspective; Third, a comprehensive implementation plan; Fourth, a demanding transformational leader. Thereon, he continues by describing three vital leadership tasks that support these attributes as follows: One, generating energy for transformation; Two, developing a vision of the future, the vision identifying a purpose and mission for the organisation, creating an emotional view for the future organisational state and providing direction to get to the vision state; Three, aligning the organisation with culture. Overall, the leadership task is creating transformation process architecture. The transformation architecture enables the transformation leader to orchestrate the transformation process. The leadership must deliberately orchestrate all the elements of the organisation's total system in order to maintain dynamic alignment and facilitate human development and organisational learning that allow forward movement without excessive risk. Failure in any one of these tasks will result in failure to transform. More importantly, these organisational theories are relevant to military institutions because they are based upon human and institutional dynamics associated with resistance, control and power present during periods of change or transformation as well as other types of institutions.

---

[2] Miles, Robert H, *Leading Corporate Transformation: A Blueprint for Business Renewal*, Jossey-Bass Inc Publishers, USA, 1997.

- **Catalysts and Theories of Military Transformation:** The catalysts for strategic military transformations have invariably been visionaries at the national apex who have provided the necessary impetus. [3]In the US, the catalyst for the transformation process commenced with former Secretary of Defense, Donald H Rumsfeld. The US Department of Defense (DoD) created US Joint Forces Command (JFC) as the transformation laboratory of the US Military to force the US Armed Forces (USAF) into jointness. [4]The Goldwater Nichols Act brought about revolutionary changes in the USAF, accelerating synergy and boosting transformation. [5]In China, change was ushered in by former Chinese premier Jiang Zemin and its implementation overseen by the Central Military Commission (CMC) and Chief of General Staff (CGS) of the PLA. In [6]Germany the transformation process was initiated by the Berlin Decree which aimed to integrate the armed forces, ensuring and reaping full benefits of ongoing technological advancements. The German Chief of the Defence Forces is overseeing the transformation of the armed forces. As for theories, [7]we may

---

[3] Garamone, Jim, 'Transforming the Military: No Magic Trick', *American Forces Press Service*, US Department of Defense, Washington, 21 June 2001, http://www.defense.gov/News/ NewsArticle.aspx?ID=45897

[4] Goldwater Nichols Department of Defense Reorganization Act of 1986, National Defense University Library, Washington DC, USA http://www.ndu.edu/library/goldnich/goldnich. html

[5] Chieh-cheng Huang, Alexander, 'Transformation and Refinement of Chinese Military Doctrine: Reflection and Critique on the PLA's View', Rand Corporation, USA http://www. rand.org/content/dam/rand/pubs/conf_proceedings/CF160/CF160.ch6.pdf

[6] Katoch, PC, 'Managing Strategic Transformation', SP's LandForces.net, http://www. spslandforces.net/story.asp?id=182

[7] 'Army Transformation: A View from the US Army War College', USA, July 2001, http:// books.google.co.in/books?id=ij8ADhCCmV4C&pg=PA135&lpg=PA135&dq=requirements +of+managing+transformation+in+us+army+as+defined+by+general+donn+a&source=bl&ot s=gm0NFRAcSo&sig=enIIObtipMVGJoSL74Cb_3lvWXY&hl=en&sa=X&ei=QcS-UY_EK YzOrQeKm4G4Cw&ved=0CCoQ6AEwAA#v=onepage&q&f=false

examine what General Donn A Starry, a former commander of US Training and Doctrine Command (TRADOC) identified when faced with the challenge of developing a doctrine to enable the US Army to defeat a much larger Soviet land force. He identified the following seven general requirements for efficient military change: One, the military leader or coalition must identify an institution or mechanism to manage change and the newly appointed institution or mechanism then must define the need for change, describing what must be done and how it differs from the present. Two, ensure that the principal staff and command personalities responsible for change possess an educational background sufficiently rigorous, demanding and relevant to bring a common cultural bias to problem solving. Three, appoint a spokesman or institution to be the champion for change. Four, build consensus for change that will give the new ideas and the need to adopt them a wider audience of converts and believers. Five, maintenance of continuity of leadership since continuity is needed among the architects of change and so that consistency of effort is brought in the process. Six, gain support from the top or near the top of the organisation on the premise that the supporter must be willing to hear out arguments for change, agree to the need, embrace the new concepts and become at least a supporter or champion for change. Seven, conduct field trials and experiments. The relevance of the proposed change must be convincingly demonstrated to a wide audience through the use of open, challenging and realistic experiments. [8] *Hope is Not a Method,* authored by General Gordon R Sullivan, former Chief of Staff US Army and Colonel Michael V Harper, deals with

---

[8] Sullivan, Gordon R and Harper, Michael V, *Hope is Not a Method,* Deli Publishing Group, Inc, New York, USA, 1996.

military transformation and lists eleven rules for guiding change: One, leading change requires leaders to do two jobs at once, in that, they must conduct today's operations and lead the organisation into tomorrow. Two, the leader uses values to signal what will not change within an organisation, and in doing so, provides stability and direction during the uncertain times associated with change. Three, the leader and his team must expend a great deal of mental effort to build a solid intellectual framework for the future. Four, the leader must change the critical process within the organisation if he wishes to effect true change as simply working on the margins and making incremental changes will not effect substantive and enduring transformation. Five, effective leaders build teams and forge alliances, as teamwork is critical to transformation and teamwork empowers people with a sense of responsibility for the organisation, thereby creating momentum for transformation. Six, resiliency and flexibility are critical for the organisation to deal with the unexpected and maintain the course throughout transformation. Seven, the transformation leader must strike a balance between resources (people, funds, time and energy) to meet today's requirement and those of tomorrow. Eight, 'better' cannot be defined using current qualitative values (better quality, reduced cycle times, shared information etc) but should include all these characteristics and more. Nine, the leader must inculcate the organisation with a positive, optimistic and creative vision of the future. Ten, accent should be on a learning organisation that learns from doing and sharing information and whose actions will spark a spirit of innovation and growth within the organisation. Eleven, creative people enable organisations in successfully transforming, hence, leaders must understand this and grow and reward creativity among all the people within the organisation.

Analysis

Many similarities exist between most theories for change. Human and institutional dynamics are generally associated with resistance, control and power. This is applicable for both military and non-military organisations and institutions. First and foremost is the need to establish a sense of urgency for transformation of the organisation. Second is the importance of leadership in successful transformation, more specifically a coalition of leaders. Third is the criticality of a vision and a strategy to achieve it. Fourth is the need for open communications to build consensus and support change. Fifth is the need to empower all people throughout the organisation to achieve the vision. And lastly, is the need to institutionalise the transformation within the organisation's culture.

The Indian military requires organisational changes that are necessary to give an impetus for synergising the armed forces towards integration and achieving unification. These changes have to be driven by the top political leadership of the country. To meet the challenge of transforming the armed forces, senior leaders and other agents of change must break the long tethers that bind the armed forces to the past. To do so they must not only compel those within the Services to alter the way they think about their traditional roles and branch missions, but also win support from the people and the nation's political leaders in their efforts to change the armed forces. However, factors like previous historical experience, a naturally conservative outlook towards change, inability to evaluate new ideas adequately and a desire by some within the organisation to preserve the status quo for fear of losing either personal or professional power within the organisation may prevent meaningful changes to occur.

The Indian Scene

Despite fighting many wars (1947, 1962, 1965, 1971, Kargil, ongoing asymmetric wars) and countering terrorism and insurgencies, we have neither been quick nor wise enough to put lessons learnt in the past to effective use. We have, unfortunately, failed in updating the defence organisations and institutional processes to foster cooperation through the required degrees of integration among the Services as well as the defence and civil Services. The reasons for such state of affairs are non existence of a permanent coordinating body for the three Services. This results in their evolution as separate entities with dissimilar establishments, wide disparity in the powers vested in the Services and the MoD, isolation of the Services from important issues concerning national security which results in their not being integrated into the system/decision-making process, and blanket ceiling on budgets, arising out of a myopic focus in defence planning, with the latter resulting in an alarming capacity gap between the Indian military and the Chinese PLA.

The Kargil War did reveal a number of failures and the GoM did make a number of recommendations. A large portion of those dealt with reform in the functioning of MoD and its interface with the armed forces. Some steps have been taken by implementing a number of recommendations. Some of which have been: Headquarters Integrated Defence Staff has been established. A Defence Intelligence Agency has been created to unify efforts of the intelligence agencies of the three Services. The Strategic Forces Command (SFC) and the newly created SFC and the Andaman & Nicobar Command as the first unified tri-Services command has been established, albeit with military assets of not much consequence. But the most essential reform of having a Chief of Defence Staff has not been implemented. The impression created about the recommendations of the Kargil Review Committee is that barring the appointment of the CDS, everything else has been implemented. However, nothing could be further from the truth as may be seen from the following: One, no CDS appointed; Two, HQ IDS created as a separate HQ instead of merging it with the MoD; Three, Chairman COSC surreptitiously eased out from controlling the SFC; Four, DIA stopped from its authorised mandate of operating transborder sources; Five, ANC remains toothless without any worthwhile military forces assigned to it; Six, Directorate General of Armed Forces Medical Services (DGAFMS) and Directorate General of Quality Assurance (DGQA) not brought under HQ IDS as envisaged. The implementation of defence reforms has, thus, been partial; more in form than in substance.

Foreign Models

It may be pertinent to study the foreign models and their experience in unification of the armed forces and management of defence. The experience gained by them and the path adopted by them may hold some lessons for us, though, we may have to be careful in implementing methods and policies adopted to suit our requirements and objectives. In post WWII era, there has been a general trend all over the world of the integration of the armed forces as well as of Services HQ with the defence ministry/DoD.

USA: For the unification of USAF, Goldwater Nichols Department of Defense Reorganisation Act of 1986 was enacted with the following main objectives: One, to recognise the DoD and strengthen civilian authority within the Department. Two, to improve the military advice provided to the president, the National Command Authority (NCA) and the National Security Council (NSC). Three, to place clear responsibility on the commanders of the unified and specified combatant commands for the accomplishment of missions assigned to those commands. Four, to increase attention to the formulation of strategy and contingency planning. Five, to provide for more efficient use of defence resources. Six, to improve joint officer management policies. Seven, to enhance effectiveness of military operations and improve the management and administration of the DoD. As mentioned, the catalysts for strategic military transformations have invariably been visionaries at the national apex who have provided the necessary impetus. In the US, the catalyst for the transformation process commenced with Donald H Rumsfeld. The US DoD created US Joint Forces Command (JFC) as the transformation laboratory of the US Military to force the USAF into jointness. The Goldwater Nichols Act accelerated synergy and boosting transformation in USAF. Through this Act, Congress clarified the command line to the combatant commanders and preserved civilian control of the military. The Act states that the operational chain of command runs from the President to the Secretary of Defense to the combatant commanders. The Act permits the President to direct that communications pass through Chairman Joint Chiefs of Staff (CJCS). This authority places CJCS in the communications chain. The US transformation was aimed towards creating a network empowered and effects oriented expeditionary force through technology combined with new concepts and organisation. The effects of US military transformation were evident during the Iraq war, where despite having fewer numbers than the Iraqi forces, the US military was able to defeat them in a short span of time. However, over time, escalating counter-insurgency operations required a new type of warfare—Human-centric Warfare.

United Kingdom: The transformation of the British military was inspired by the US military and their ideas on changing the character of warfare. In the UK, the debate over the CDS raged for eighteen years and then the government simply forced the CDS on the Services. The British military adopted a slightly different version of the US NCW called Network Enabled Capability (NEC), which focuses more on the human component. The British recognise that future operations will be characterised by Effects Based Approach to Operations which will involve a combination of military and non-military instruments to achieve the end result. The changing nature of conflict has forced military forces not only to fight wars, but also establish the groundwork for sustained governance. The British military in Afghanistan applied EBAO by using military instruments to achieve non-military tasks that they termed as soft effects. In Afghanistan, the British military eventually established effective communication with the local population, based on cultural understanding, in order to win over their support.

Germany: Reform of the Bundeswehr (German armed forces) had been debated for decades but the political establishment had always proved resistant to change. The transformation process was finally initiated through the Berlin Decree which aimed to integrate the armed forces, ensuring and reaping full benefits of ongoing technological advancements. The German Chief of the Defence Forces is overseeing the transformation of the armed forces. There has been considerable downsizing with elimination of conscription and modernisation. Post the experience in Kosovo, the Bundeswehr focused on acquiring new skills, adapting to new, fast-changing threats and enemies including international terrorism cells, combined with the need for structural reforms. The Minister of Defense or the Chancellor is supported by the Chief of Defense (CHOD) and the service chiefs in his function as Commander-in-Chief. The CHOD and the service chiefs form the Military Command Council with functions similar to those of the Joint Chiefs of Staff in the US. Subordinate to the CHOD is the Armed Forces Operational Command. For smaller missions one of the Service HQ may exercise command and control of forces in missions abroad. Generally speaking, the Bundeswehr is still among the world's most technologically advanced and best-supplied militaries, as befits Germany's overall economic prosperity and infrastructure.

China: With the total grip of the Chinese Communist Party on the country including on the PLA, which is fully integrated into strategic decision-making and all matters military, China needed no Act of Parliament to usher in transformation including jointness in the Army. However, China pointedly drew lessons from hi-tech wars by the US and NATO forces, and accelerated transformation was ushered in by former premier Jiang Zemin. Its implementation was overseen by the Central Military Commission (CMC) and Chief of General Staff (CGS) of the PLA. The Chinese security apparatus is made up of the Ministry of State Security and the Ministry of Public Security, the People's Armed Police, PLA and the state judicial and penal system. The PLA remains central to China's security. The official defence budget of US $90 bn (unofficially to the tune of US $150 bn) is likely to be doubled, if western media reports are to be believed.

Requirement

While it is generally acknowledged that in India a CDS is yet to be appointed, jointness in Services simply has to be forced from the top and perhaps only an Act of Parliament can enforce this. We need to focus on a template for gaining irreversible momentum that centres on the following six tasks: One, instill a sense of urgency for change within and external to the organisation; Two, establish a powerful coalition to guide the change; Three, develop a vision for the future and a strategy to achieve it; Four, communicate the vision and the need for change; Five, empower people within the organisation to include rewarding imagination and creativity; Six, institutionalise the change within the organisation.

The reason why India has chosen to maintain status quo since independence, is not because the need has not been felt, but because of sectional interests that have little to do with the security of the nation. The price for military complacency that we must pay will never be really known till war is thrust upon us. There is an urgent need for change in the higher defence organisation of the country. Hence, the requirement for immediate reforms in our higher defence organisation that can broadly be enunciated as follows:

- Completed integration of HQ IDS with the MoD.
- Integration of Services HQ with MoD to promote a much greater role for the armed forces in the decision-making process related to defence and national security issues. Mere renaming of Services HQ as IHQ of MoD (Army/Navy/Air Force) is mere lip service.
- Restructuring of the NSC to include the CDS (once appointed) as a permanent member.
- Delineation of the National Security Advisor and deputy National Security Advisor in secondary roles of secretary and additional secretary to the Prime Minister respectively with both these posts by rotation held by experienced military officers and civil servants, in addition to a constitutional NSAB with a full-time staff to be effective in its role.
- The appointment of a five-star officer as CDS who would be the principal military advisor to the Prime Minister in addition to being the head of the Joint Chiefs of Staff and charged with issuing instructions for joint operations and exercising control over joint organisations and strategic forces.
- Creation of a permanent Joint HQ under the CDS with adequate staff and infrastructure.
- Enactment of an Act of Parliament for transformation of Indian armed forces—draft to be prepared by the National Security Council (NSC).
- Development of a robust C^4I^2SR structure in the higher

defence organisation with the focus on a joint strategy on Information Warfare for better battlespace management.

- Establishment of Integrated Theatre Commands (ITCs) and Integrated Functional Commands (IFCs).

The process of integration of the three Services has started taking shape with establishment of HQ IDS, SFC, ANC etc. post the Kargil Review Committee recommendations. However, there is a need to firm up. The suggested road map to achieve joint military capabilities is as follows: One, appointment of a CDS and establishment of a Joint HQ by 2014; Two, integrated HQ IDS fully with MoD by 2015; Three, enactment of an Act of Parliament for Transformation of Indian armed forces by 2015—draft to be prepared by the NSC; Four, creation of ITCs and ILCs including an Integrated Aero-Space Command, Integrated Special Forces Command, Integrated Logistics Command(s) and Integrated Training Command(s) and the like by 2020; Five, establishment of a robust C^4I^2SR structure in the higher defence organisation by 2020.

It goes without saying that the Services chiefs themselves too are a hindrance to transformation and jointness. A former Army Chief who was a proponent for immediate appointment of a CDS with 'full operational powers' did a turnaround just prior to superannuation by saying that we should not go for a CDS till the time we do not sort out our insurgencies, as if the two are related. Ironically, the same former COAS had stated during a Unified Commanders Conference, 'We have very good synergy amongst the Services chiefs because we get together once a month and then have breakfast together.' In a more recent jolt to jointness, the three Services chiefs have ruled out formation of Theatre Commands on the ground that these would eat into their individual turfs. In the UK, the debate over the CDS raged for eighteen long years with Services chiefs haggling over individual turfs and this is precisely what is happening in

India today. However, in the UK, the political hierarchy simply thrust the CDS down the throat of the military. Former Army Chief, General S Padmanabhan had gone on record to say, 'There is no escaping the military logic of creating suitably constituted Integrated Theatre Commands (ITCs) and Integrated Functional Commands (IFCs) as a whole.' Similarly, former Army Chief, General VP Malik had said, 'It is not my case that the Services chiefs do not cooperate in war. Were they not to do so, it would be churlish. But in war, cooperative synergies are simply not good enough.' More significantly, Prime Minister Manmohan Singh, addressing the Unified Commanders Conference in 2004 had stated, 'Reforms within the armed forces also involve recognition of the fact that our Navy, Air Force and Army can no longer function in compartments with exclusive chains of command and single service operational plans'. Similar to the UK, the political authority in India needs to thrust the CDS, ITCs and IFCs down the throat of the military. This is all the more required considering the existing multiple threats to national security.

National security imperatives demand that integration and reorganisation of the MoD, Service HQ and the armed forces be progressed in a phased manner after careful consultation and planning since major attitudinal and work culture reorientation is involved and elaborate preparatory and education work will be necessary. The changes will be fundamental and initially disruptive in nature, and should be effected without significantly compromising force readiness. The main objectives should be greater force effectiveness, organisational efficiency, overall economy and personnel reduction through amalgamation of functions and elimination of unnecessary ones. The needs and benefits of innovation and change have to be clearly demonstrated to the military organisation. The changes may threaten traditions and methods dear to many but their hearts and minds need to be won over for a successful change in the armed forces. For change to occur, senior leaders must fight through strength of logic and will to gain a consensus for the future. For the armed forces to succeed, all existing communications channels to open dialogue for transformation must be utilised. Manifestations of change are generally resisted at all levels particularly in a country like India. The military is as rigid, if not more, especially considering that transformation requires much more than mere change, the latter being slow moving and incremental. Take the example of the fact that it is the creative people who enable organisations to successfully transform, which leaders must understand, encourage its growth and reward creativity among all members of the organisation. However, what is perplexing to the military is that this rule is hard to support given the tension between adherence to discipline and the questioning attitude required for innovative and creative thought within the military.

16

The Future

Network-centric Warfare enables a shift from attrition-style warfare to a much faster and more effective warfighting style characterised by the new concepts of speed of command and self-synchronisation. The need to transform the military into an NCW capable force requires accelerated focus but at the same time we should not lose sight of the fact that NCW is a mere enabler for EBOs which is 'the' requirement of present and future threats. And, EBOs require a synergised effort of the entire nation, not just the military.

Having established NCW as a critical area of military exploration and application, the natural follow-up activity should be the total transformation of the military. EBO is neither a new form of operations nor is it new. It is actually a way of thinking or a methodology for planning, executing and assessing operations designed to attain specific effects that are required to achieve desired national security objectives. To this end, military transformation will enable the Indian military to achieve broad and sustained competitive advantage in the twenty-first century. It comprises of

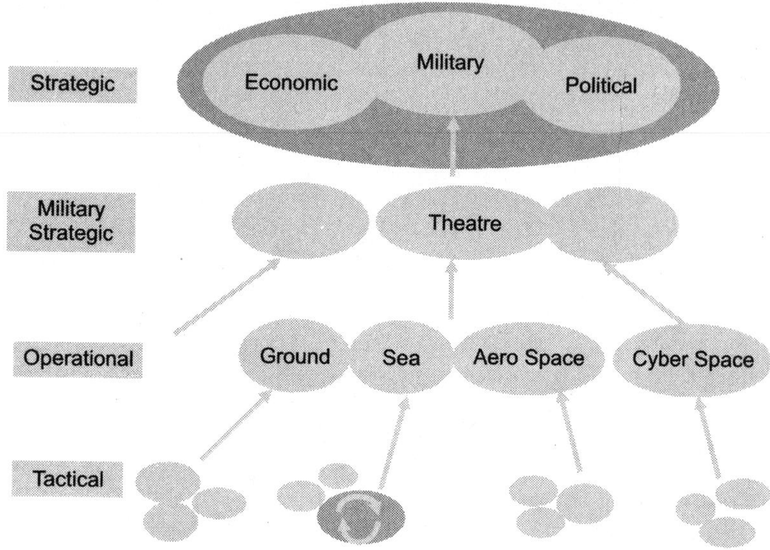

those activities that anticipate and create the future by co-evolving concepts, processes, organisations and technologies to produce new sources of military power. The transformation of our armed forces will dramatically increase our strategic and operational responsiveness, speed, reach and effectiveness, making our forces increasingly precise, lethal; adjustable, agile, survivable and more easily sustainable. Essentially, transformation is the shaping and moulding of our military force that seeks to fully exploit the advantages we currently possess and to protect against and minimise our vulnerabilities. Transformation is employed and accomplished through a combination of concepts, capabilities, people and technology. The overall objective of these changes should be to develop and retain competitive advantage in warfare over our adversaries. In this process of transformation, our focus must remain on forward deterrence in all domains of war and equally importantly, within the spectrum of conflict in the sub-conventional space.

Epilogue

[1]Former Chief of Army Staff, General S Padmanabhan wrote in his maiden book about a fictitious futuristic scenario of India checkmating America through cyber warfare and electromagnetic pulse attacks. The assumption of such a scenario was based on multiple premises including continued backing of Pakistan at the cost of India, US policy of pre-emption, demonstrated hegemony at international forums including at the UN and more significantly, economic forums with crippling effects on emerging economies. The evolving scenario in this book, *The Writing on the Wall: India Checkmates America, 2017*, shows China dumping Pakistan and signing a defence treaty with India, on the lines of the Indo-Soviet Treaty of 1971. Consequently, with increasing threat to the mainland, the US signs a defence pact with Pakistan and in its desperation in preventing an Asian Security Architecture centred on India-China defence pact, special operations components from US Pacific and Central Commands are deployed in South Asia, in league with the Central Intelligence Agency (CIA) and the ISI, to keep India busy, which eventually leads to a confrontation with India checkmating the US.

---

[1]Padmanabhan, S, General, *The Writing on the Wall – India Checkmates America* 2017, Manas Publications, New Delhi, India, 2004.

At the time of publication of the book in 2004, there were criticisms on two counts: One, as to how can there be confrontation between India and the US, considering relations between these two countries, and how can India ever take on America—David versus Goliath. But critics were wrong on both counts. First, there are no permanent friends or enemies where national interests are concerned—America's own example being foremost. Then look at Russia-China relations, who have fought war but today, are friends and linked economically and have strong military–industrial relations. But that does not imply that they will remain friends forever. By all indications they are headed for a confrontation at a future date, be it China's silent demographic invasion in Russia's Far East (already estimated to be some 100 mn), clash for control of Central Asia's natural resources and the like. Second, is the David versus Goliath. Well, David did have his slingshot but today, asymmetries can actually be balanced out by focused investments in the cyber, space and electromagnetic domains of war, aside from nuclear power. That is where China is desperately seeking to enhance capabilities to compensate for superior military power. That is how China plans to stop American Carrier Battle Groups from reaching the conflict zone in the South China sea to assist Taiwan.

Whether India will need to confront the US in such a manner is a matter of conjecture but such a threat from China, as also China–Pakistan combination, is very real. India would do well to prepare for checkmating both China and Pakistan.

Bibliography

Books

1. Alberts, David S and Hayes, Richard E, *Power to the Edge, Command and Control Research Program*, Center for Advanced Concepts and Technology (ACT), Department of Defense, USA, 2005.

2. Bailey, Alvin L, *The Implications of Network Centric Warfare*, US Army War College Research Project, USA, 2004.

3. Basalla, George, *The Evolution of Technology*, Cambridge University Press, USA, 1988.

4. Boot, Max, *War Made New: Technology, Warfare, and the Course of History: 1500 to Today*, Gotham Books, USA, 2006.

5. Bowden, Mark, *Black Hawk Down: A Story of Modern War*, Penguin Books, UK, 2000.

6. Brenner, Susan W, *Cyber Threats: The Emerging Fault Lines of the Nation State*, Oxford University Press, USA, 2009.

7. Chan, Serena, *Architectures for a Space-Based Information Network with Shared On-orbit Processing*, Massachusetts Institute of Technology, USA, 2005.

8. Cotterell, Arthur, *Chariot: The Astounding Rise and Fall of The World's First Flying Machine*, Pimlico, UK, 2004.

9. Creveld, Martin van, *Technology and War: From 2000 BC to the Present*, Free Press, USA, 1991.

10. Dayal, Shantanu, *C4I2SR for Indian Army*, Bibliophile South Asia, India, 2012.

11. Elangovan, K, *GIS: Fundamentals, Applications and Implementations*, New India Publishing Agency, India, 2006.

12. Friedel, Robert, *A Culture of Improvement: Technology and the Western Millennium*, MIT Press, UK, 2007.

13. Gaddis, John Lewis, *The Long Peace: Inquiries Into the History of The Cold War*, Oxford University Press, UK, 1989.

14. Gould, Michael and Hecht, Louis, Jr, *OGC: A Framework for Geospatial and Statistical Information Integration*, Open GIS Consortium, OGC White Paper, Estonia, 2001.

15. Gray, Collin, *Strategy for Chaos—Revolution in Military Affairs and the Evidence of History*, Crown House, UK, 2005.

16. Jayaswal, Vinod Kumar, *Cyber Crime & Cyber Terrorism*, RVS Books, India, 2010.

17. Kaplan, Robert D, *The Revenge of Geography*, Random House, USA, 2012.

18. Katoch, PC and Datta, Saikat, *India's Special Forces – History and Future of Indian Special Forces*, Vij Books India Pvt Ltd, India, 2013.

19. Kotter, John P, *Leading Change*, Harvard Business Review Press, USA, 2012.

20. Lawrence, Philip K, *Modernity and War: The Creed of Absolute Violence*, Martin's Press Inc., USA, 1997.

21. Liang, Qiao and Xiangsui, Wang, *Unrestricted Warfare*, PLA Literature and Arts, Publishing House, China, 1999.

22. McNeill, William H, *The Rise of the West: A History of the Human Community*, University of Chicago Press, USA, 1991.

23. Miles, Robert H, *Leading Corporate Transformation: A Blueprint for Business Renewal*, Jossey-Bass Inc Publishers, USA, 1997.

24. O'Connell, Robert L, *Of Arms and Men: A History of Wars, Weapons and Aggression*, Oxford University Press, USA, 1989.

25. Oberoi, Vijay, Lt Gen, *Net-Centric Warfare*, Centre for Land Warfare Studies, KW Publishers Pvt Ltd, India, 2008.

26. Padmanabhan, S, General, *The Writing on the Wall – India Checkmates America 2017*, Manas Publications, India, 2004.

27. Power, Daniel J, *Decision Support Systems: Concepts and Resources for Managers*, Westport, Conn., Quorum Books, USA, 2002.

28. Roy, Rana Deb, *The Great Trignometrical Survey of India in Historical Perspective*, Indian Journal of Science, 21(1): 22-32, 1986.

29. Sawyer, Ralph, and Sawyer, Mei-chün Lee, *The Art of War*, Westview Press, 1994.

30. Sullivan, Gordon R and Harper, Michael V, *Hope is Not a Method*, Deli Publishing Group, Inc, USA, 1996.

31. Thornton, Rod, *Asymmetric Warfare: Threat and Response in the Twenty-First Century*, Polity Press, USA, 2007.

32. Toffler, Alvin and Toffler, Heidi, *War and Anti-War: Survival at the Dawn of 21st Century*, Warner Books, USA, 1995.

33. Turban, Efraim; Aronson, Jay E and Liang, Ting-Peng, *Decision Support Systems and Intelligent Systems*, Pearson-Prentice Hall, USA, 2008.

Articles/Web Reference

1. 'Accelerate your transformation to the software-defined data center in the world's most powerful, trusted, and smart family of storage products', *EMC*, http://www.emc.com/storage/symmetrix-vmax/symmetrix-vmax.htm

2. Ahvenainen, Sakari, 'Backgrounds and Principles of Network-centric Warfare', National Defence College, November 2003, http://www.carlisle.army.mil/dime/documents/NCW%20Background%20Principles.pdf

3. Sieff, Martin, 'US Space Assets Vital but Vulnerable', 2005, http://www.spacewar.com/news/milspace-05x.html

4. Alberts, David S; Garstka, John J and Stein, Fredrick P, 'Net Centric Warfare: Developing and Leveraging Information Superiority', Department of Defense C4ISR Cooperative Research Program, http://www.dodccrp.org/files/Alberts_NCW.pdf

5. Ali, Irena, 'Is NCW Information Sharing a Double Edged Sword — Voices from the Battlespace', 2006, http://www.dsto.defence.gov.au/attachments/Ali_Is%20NCW%20Information%20Sharing%20a%20Double%20Edged%20Sword%20-%20Voices%20from%20the%20Battlespace.pdf

6.   Alspaugh, Chris; Davé, Nikhil; Hepner, Tom; Leidy, Andy and Stell, Mark, 'Modeling and Simulation in Support of Network Centric Warfare Analysis', Command and Control Research and Technology Symposium (CCRTS), 2004, http://www.dtic.mil/cgi-bin/GetTRDoc?AD=ADA466029

7.   Anand, Vinod, 'Joint Development of Inter-Services Network and C4I2 Systems', Institute for Defence Studies and Analyses, 2009, http://www.idsa-india.org/an-oct-00-9.html

8.   'Army Transformation: A View from the US Army War College', 2001, http://books.google.co.in/books?id=ij8ADhCCmV4C&pg=PA135&lpg=PA135&dq=requirements+of+managing+transformation+in+us+army+as+defined+by+general+donn+a&source=bl&ots=gm0NFRAcSo&sig=enIIObtipMVGJoSL74Cb_3lvWXY&hl=en&sa=X&ei=QcS-UY_EKYzOrQeKm4G4Cw&ved=0CCoQ6AEwAA#v=onepage&q&f=false

9.   Banned Terrorist Organizations, Schedule I – First Schedule (of the UA(P) Act, 1967), http://www.nia.gov.in/banned_org.aspx

10. Bernard, Simon, 'The Revolution in Military Affairs: Approach with Caution', *The Army Doctrine and Training Bulletin*, Volume 3, No 4/Volume 4, No 1, http://www.army.forces.gc.ca/caj/documents/vol_03/iss_4/CAJ_vol3.4_13_e.pdf

11. Blash, Edmund C, 'Network-centric Warfare Requires A Closer Look', IWS – The Information Warfare Site, 2003, http://www.iwar.org.uk/rma/resources/ncw/ncw-forum.htm

12. Carr, Peter, Information Technology and Modern Warfare 2013 Update, 2013, http://impactofinformationsystemsonsociety.wordpress.com/2013/02/28/information-technology-and-modern-warfare-2013-update/

13. Catanzaro, Jean, 'User-Centered Design Requirements for Maritime Chat Clients in Network-centric Warfare', 2006, http://www.dsto.defence.gov.au/attachments/Catanzaro%20&%20Smillie_User-Centered%20Design%20Requirements%20for%20Maritime%20Chat%20Clients%20in%20Network-Centric%20Warfare.pdf

14. Chandele, AKS, Lt Gen, Editorial, *GEO Intelligence*, Vol 3, Issue 3, 2013, India.

15. Chapman, Gary, 'An Introduction to the Revolution in Military Affairs', 2003, http://www.lincei.it/rapporti/amaldi/papers/XV-Chapman.pdf

16. Charney, Scott and Werne, Eric T, 'Cyber Supply Chain Risk Management: Toward a Global Vision of Transparency and Trust', Microsoft Corporation, 2011, http://www.microsoft.com/en-us/download/confirmation.aspx?id=26826

17. Chieh-cheng Huang, Alexander, 'Transformation and Refinement of Chinese Military Doctrine: Reflection and Critique on the PLA's View', Rand Corporation, http://www.rand.org/content/dam/rand/pubs/conf_proceedings/CF160/CF160.ch6.pdf

18. 'China deploys 11,000 troops in Gilgit area in Pakistan occupied Kashmir', *Daily New Analysis*, 2010, http://www.dnaindia.com/india/1430066/report-china-deploys-11000-troops-in-gilgit-area-in-pakistan-occupied-kashmir

19. 'Chinese Military Modernisation and Force Development: A Western Perspective', Centre for Strategic & International Studies, http://csis.org/publication/chinese-military-modernization-and-force-development-0

20. Christian, OG, 'Understanding the Prussian-German General Staff System', 1992, http://www.dtic.mil/dtic/tr/fulltext/u2/a249255.pdf

21. 'Coordinate transformation from WGS 1984 to Everest 1830 or vice versa', *ArcGIS Resources*, http://forums.arcgis.com/threads/53532-Coordinate-transformation-from-WGS-1984-to-Everest-1830-or-vice-versa

22. Cowen, Michael B and Fleming, Robert A, 'Collaboration in Command and Control Environments: Exchanging Iconic Tags of Key Information', 2006, http://www.dsto.defence.gov.au/attachments/Cowan%20&%20Fleming_Collaboration%20in%20Command%20and%20Control%20Environments%20-%20Exchanging%20Iconic%20Tags%20of%20Key%20Information.pdf

23. Custer, John M, 'Small devices and big data', *Armed Forces Journal*, 2012, http://www.armedforcesjournal.com/2012/10/11458170/

24. DEF SIM 2012, Military Modeling and Simulation Solutions Seminar, 2012, http://www.edstechnologies.com/Mailer/aug12/news_3.html

25. Dekker, Anthony H, 'Revisiting 'SCUDHunt' and the Human Dimension of NCW: Some Thoughts', 2006, http://www.dsto.defence.gov.au/attachments/Dekker_SCUDHunt.TTCP.pdf

26. 'DoD Military Lessons Learned – Joint Army, Air Force, Navy, Marines', http://www.au.af.mil/au/awc/awcgate/awc-lesn.htm

27. Dolan, Nancy and Narkevicius, Jennifer McGovern, 'Systems Engineering, Acquisition and Personnel Integration (SEAPRINT): Achieving the Promise of Human Systems Integration', 2006, http://ftp.rta.nato.int/public/PubFullText/RTO/MP/RTO-MP-HFM-124/MP-HFM-124-01.pdf

28. Dubey, Muchkund, 'India facing threats from almost all its neighbours', *Indian Express*, 02 September 2012, http://www.indianexpress.com/news/india-facing-threats-from-almost-all-its-neighbours-says-former-indian-foreign-secy/996696/

29. 'Emerging Threats in the 21st Century-Strategic Foresight and Warning', Center for Security Studies (CSS), 2007, http://www.isn.ethz.ch/Digital-Library/Publications/Detail/?lng=en&id=47160

30. 'Empowering the Army', Army Modeling and Simulation Office, US Army, http://www.ms.army.mil/

31. Floyd D Spence 'National Defense Authorization Act for Fiscal Year 2001 – Report of the Committee on Armed Forces', http://asafm.army.mil/Documents/OtherDocuments/CongInfo/BLDL/HR//01AUTHh.pdf

32. Fogarty, Gerard, Brigadier, 'Progressing the Human Dimension of NCW in the ADF', 2006, http://www.dsto.defence.gov.au/attachments/Keynote%20Address%20Brigadier%20Fogarty_Progressing%20the%20Human%20Dimension%20of%20NCW%20in%20the%20ADF.pdf

33. Garamone, Jim, 'Transforming the Military: No Magic Trick', *American Forces Press Service*, US Department of Defense, 2001, http://www. defense.gov/News/NewsArticle.aspx?ID=45897

34. 'Gartner Says India IT Infrastructure Spending Will Reach $2.3 Billion By 2014', *Gartner*, http://www.gartner.com/newsroom/id/2481916

35. *GEO Intelligence*, Vol 3, Issue 3, 2013, pp 6-7.

36. 'Geospatial Defense and Intelligence in Afghanistan', 6th Annual Geospatial Intelligence Middle East 2013, GIS for Defense, Intelligence & Security, http://www.geospatialdefence.com/redForms.aspx?eventi d=7079&id=389080&FormID=%2011&frmType=Additional%20 Content&m=4888&FrmBypass=False&mLoc=U&SponsorOpt=False

37. 'Geospatial Intelligence in Joint Operations', *Joint Publication* 2-03, US Army, 2012, http://www.dtic.mil/doctrine/new_pubs/jp2_03.pdf

38. Goldwater Nichols Department of Defense Reorganization Act of 1986, National Defense University Library, http://www.ndu.edu/ library/goldnich/goldnich.html

39. Gray, Collin S, 'Security Threats in the 21st Century', University of Reading, 2006, http://www.dtic.mil/cgi-bin/GetTRDoc?AD= ADA481210

40. Guha, Manabrata, 'China and the RMA Today', *ISN*, 2013, http:// www.isn.ethz.ch/Digital-Library/Articles/Special-Feature/Detail/?lng =en&id=162697&contextid774=162697&contextid775=162694& tabid=1454241796

41. Hamlen, Kevin; Kantarcioglu, Murat; Khan, Latifur and Thuraisingham, Bhavani, 'Security Issues for Cloud Computing', *International Journal of Information Security and Privacy*, 4(2), 39-51, 2010 39.

42. Hoffman, Frank G, 'Hybrid Threats: Reconceptualizing the Evolving Character of Modern Conflict', http://www.ndu.edu/inss/ docuploaded/StrategicForum_240.pdf

43. Honabarger, Jason B, BS, 'Modeling Network Centric Warfare (NCW) with the System Effectiveness Analysis Simulation (SEAS)', Air Force Institute of Technology, Ohio, 2006, http://www.dtic.mil/ cgi-bin/GetTRDoc?AD=ADA446395

44. Human Factors in Network Warfare Symposium, Australian Government, Department of Defence, Defence Science and Technology Organization, May 2006, http://www.dsto.defence.gov.au/events/4659/

45. Huth, Alexa and Cebula, James, 'The Basics of Cloud Computing', *US-CERT*, http://www.us-cert.gov/sites/default/files/publications/CloudComputingHuthCebula.pdf

46. Ibrugger, Lothar, 'The Revolution in Military Affairs', NATO Parliamentary Assembly, 1998, http://www.iwar.org.uk/rma/resources/nato/ar299stc-e.html

47. 'India's National Map Policy: HOPE vs HYPE', *Coordinates*, 2005, http://mycoordinates.org/india%E2%80%99s-national-map-policy-hope-vs-hype/all/1/

48. 'Indian security forces discover 'unknown hill' using g-tech', *Geospatial World*, 2012, http://www.geospatialworld.net/News/View.aspx?ID=24574_Article

49. 'Information Assurance for Allied Communications and Information Systems', Combined Communications Electronics Board (CCEB), ACP (F), NATO, 2008, http://jcs.dtic.mil/j6/cceb/acps/acp122/ACP122F.pdf

50. 'Information Management for Net-Centric Operations', Defence Science Board Summer Study, Department of Defense, 2006, http://www.acq.osd.mil/dsb/reports/ADA467538.pdf

51. 'Instrument Set, Reconnaissance and Surveying (ENFIRE)', Northrop Grumman, http://www.northropgrumman.com/Capabilities/ENFIRE/Documents/ENFIRE.pdf

52. Janu, Raveen, 'F-INSAS: India's Future Soldier', Centre for Land Warfare Studies, 2012, http://www.claws.in/index.php?action=master&task=1219&u_id=187

53. Janu, Raveen, 'India's Network Centric Warfare Programme-Tactical Communication System Part II', Centre for Land Warfare Studies, 2013, file:///C:/Users/tara/Documents/TCS%20Part%20II.htm

54. Ji, You, 'The Revolution in Military Affairs and the Evolution of China's strategic Thinking', *Contemporary Southeast Asia*, Volume 21,

1999, http://www.jstor.org/discover/10.2307/25798464?uid=3738 256&uid=2129&uid=2&uid=70&uid=4&sid=21102365801387

55. Joint Lessons Learned Program, Chairman of The Joint Chiefs of Staff Instruction J-7, CJCSI3150.25E, US Military, 2012, http://www.dtic.mil/doctrine/doctrine/jllpb/cjcsi3150_25.pdf

56. *Journal of the Asiatic Society of Bengal*, Internet Archive, http://archive.org/stream/journalofasiati291860asia/journalofasiati 291860asia_djvu.txt

57. Kak, Kapil, 'Revolution in Military Affairs—An Appraisal', Institute for Defence Studies and Analyses, http://www.idsa-india.org/an-apr-01.html

58. Kamradt, Henry and MacDonald, Douglas, 'The Implications of Network Centric Warfare for United States and Multinational Military Operations', Occasional Paper, Center for Naval Warfare Studies, http://www.dtic.mil/cgi-bin/GetTRDoc?AD=ADA430553

59. Kanwal, Gurmeet, 'The Imperative of Modernising Military Communication Systems', Institute for Defence Studies and Analyses, 2010, http://www.idsa.in/idsacomments/TheImperativeof ModernisingMilitaryCommunicationsSystems_gkanwal_160210

60. Katoch, PC, 'Light beyond the Tunnel', *SP's MAI*, Vol 3, Issue 13, SP Guide Publications.

61. Katoch, PC, 'Managing Strategic Transformation', SP's LandForces. net, http://www.spslandforces.net/story.asp?id=182

62. Katoch, PC, 'Special Operations', United Services Institution of India, 2013, file:///C:/Users/tara/Documents/All%20Articles%20 Published/USI%20-%20Special%20Operations.htm

63. Katoch, PC, 'Surreal China – Lessons Nextdoor', Centre for Land warfare Studies, 2013, http://www.claws.in/Surreal-China---Lessons-Nextdoor-Prakash-Katoch.html

64. Katoch, PC, 'Transform National security Apparatus', SP's LandForces.net, http://www.spslandforces.net/story.asp?id=140

65. Klonoski, Jake, 'Fighting Smart wars in Smart Ways', *Oregonlive*, 2011, http://www.oregonlive.com/opinion/index.ssf/2011/08/fighting_ smart_wars_in_smart_w.html

66. Krishnamurthy, J; Natrajan, S; Krishnaiah, P; Kalyanramn, K; Rao, Mukund and Jayaramn, V, 'Global Spatial data Infrastructure: Large Scale Mapping using high resolution satellite data', http://www.gsdidocs.org/gsdiconf/GSDI-7/papers/TSsdSN.pdf

67. Kuhn, James K, 'Network Centric Warfare: The End of Objective Oriented Command and Control', Naval War College, 1998, http://www.dtic.mil/dtic/tr/fulltext/u2/a348447.pdf

68. Lambert Conformal Conic, http://www.georeference.org/doc/lambert_conformal_conic.htm

69. Lawlor, Maryann, 'Military Aims to Cache in on Stored Data', *Signal Magazine*, 2001, http://www.afcea.org/signal/subjectindex/software.html

70. Lawson, Shannon M, 'Information Warfare: An Analysis of the Threat of Cyber Terrorism Towards the US Critical Infrastructure', Global Information Assurance Certification Paper, 2005, http://www.giac.org/paper/gsec/1621/information-warfare-analysis-threat-cyberterrorism-us-critical-infrastr/102978

71. Lister, Jonathan, 'The Impact of Technology on Warfare', 2013, http://www.ehow.com/info_7844707_impact-technology-warfare.html

72. Maria, Anu, 'Introduction to Modeling and Simulation', 1997, http://www.inf.utfsm.cl/~hallende/download/Simul-2-2002/Introduction_to_Modeling_and_Simulation.pdf

73. Mazarr, Michael J, 'The Revolution in Military Affairs: A framework for Defense Planning', Institute for Strategic Studies, 1994, http://www.strategicstudiesinstitute.army.mil/pdffiles/pub242.pdf

74. Metcalfe's Law, http://www.princeton.edu/~achaney/tmve/wiki100k/docs/Metcalfe_s_law.html

75. Military Simulation & Training India, 2012, http://www.shephardmedia.com/static/files/download/MSTI-Brochure-2012.pdf

76. Modeling and Simulation (M&S): Associations/Organizations/Committees, http://www.site.uottawa.ca/~oren/links-MS-AG.htm

77. 'Modern information technologies to simulate use of space-based assets', http://law-journals-books.vlex.com/vid/technologies-simulate-space-assets-56653967

78. Nagal, BS, Lt Gen, 'Space: The New Battleground', *GEO Intelligence*, Vol 3, Issue 3, 2013, India.

79. Naing, Saw Yan, 'China Sells Helicopter Gunships to USWA: Report', *The Irrawaddy*, 2013, http://www.irrawaddy.org/archives/33350

80. Nanavatty, RK, Lt Gen, 'Changing Nature of Conflict: Trends and Responses', *Manekshaw Paper No 18*, 2010, Centre for Land Warfare Studies, http://www.claws.in/administrator/uploaded_files/1267073084MP_18.pdf

81. Narkevicius, Jennifer McGovern and Owen, John E, 'Human Systems Integration Enhancing Performance in Network Centric Warfare: The SEAPRINT Perspective', USA, 2006 http://www.dsto.defence.gov.au/attachments/Narkevicius,%20Stark%20&%20Owen_Pending%20Official%20Release_Human%20Systems%20Integration%20Enhancing%20Performance%20in%20Network%20Centric%20Warfare.pdf

82. National Geospatial-Intelligence Agency, https://www1.nga.mil/Pages/default.aspx

83. Nimmo, Kurt, 'The Real Benghazi Story: US Op to Arm al-Qaeda in Syria', 2013, http://www.infowars.com/the-real-benghazi-story-us-op-to-arm-al-qaeda-in-syria/

84. Patrick, Charles, 'India: Future of Change – Will the Greatest Future Security Threats to India be from Inside or Outside the Country?', 2011, http://www.indiafutureofchange.com/featureessay_d0091.htm

85. Pester-DeWan, Joanne and Oonk, Heather, 'Human Performance Benefits of Standard Measures and Metrics for Network-centric Warfare', 2006, http://www.dsto.defence.gov.au/attachments/Pester-DeWan,%20Oonk%20&%20Smillie_Human%20Performance%20Benefits%20of%20Standard%20Measures%20and%20Metrics%20for%20Network-Centric%20Warfare.pdf

86. Pironti, John P, 'Key Elements of an Information Security Program', Information Systems Audit and Control Association, 2005, http://www.isaca.org/Journal/Past-Issues/2005/Volume-1/Documents/jpdf051-Key-Elements-of-an-is-program.pdf

87. Plain English ISO IEC 27002 2005: Information Security Control Objectives, http://www.praxiom.com/iso-17799-objectives.htm

88. Poirier, John A; Bates, Edgar and Tempstli, Mark, 'How Much is a Pound of C4ISR Worth-An Assessment Methodology to Evolve Network Centric Measures and Metrics for Application to FORCEnet', http://www.dodccrp.org/events/8th_ICCRTS/pdf/143.pdf

89. Power, DJ, 'What is a DSS?', *The On-Line Executive Journal for Data-Intensive Decision Support 1(3)*, 1996, http://dssresources.com/papers/dssarticles.html

90. Reddy, Sridhar S, 'Emerging Trends in Data Center Industry: A Ctrl S Perspective', *Ctrl S*, http://www.ctrls.in/downloads/emerging_data_center_trends.pdf

91. Remote Sensing Data Policy (RSDP – 2011), http://www.isro.org/news/pdf/RSDP-2011.pdf

92. Renner, Scott, 'Net-Centric Information Management', 2005, http://www.dodccrp.org/events/2005/10th/CD/papers/348.pdf

93. Riesterer, Jim, 'UTM – Universal Transverse Mercator Geographic Coordinate System', http://geology.isu.edu/geostac/Field_Exercise/topomaps/utm.htm

94. Roland, Alex, 'War and Technology', Foreign Policy Research Institute, Vol 14, No 2, 2009, http://www.fpri.org/footnotes/1402.200902.roland.wartechnology.html

95. Sahgal, Arun and Anand, Vinod, 'Revolution in Military Affairs and Jointness', *Journal of Defence Studies*, Volume 1, No 1, Institute for Defence Studies and Analyses, http://idsa.in/system/files/jds_1_1_asahgal_vanand.pdf

96. Sehgal, Arun, Brig, 'Militarisation of Space: India Centric Thoughts and Perspectives', *GEO Intelligence*, Vol 3, Issue 3, 2013.

97. Sharma, Abhimanyu, 'Military Training & Simulation: Part II', *South Asia defence & Strategic Review*, 2012, http://www.defstrat.com/exec/frmArticleDetails.aspx?DID=375

98. Sharoni, Asher H and Bacon, Lawrenica D, 'The Future Combat System (FCS): A Satellite-fueled, Solar-Powered Tank?' *Armor*, 1998.

99. Wu, Jun; Xiangli, Sun and Side, Hu, 'The Impact of Revolution in Military Affairs on China's Defense Policy', Institute of Applied Physics and Computational Mathematics, 2003, http://www.lincei.it/rapporti/amaldi/papers/XV-WuRMAImpact.pdf

100. Singh, Mandeep, 'Asymmetric Wars In The Indian Context', Institute of Defence Sudies and Analyses, 2011, http://www.idsa.in/idsacomments/AsymmetricWarsintheIndianContext_msingh_131011

101. Sishtla, Srinivas, 'Indian Military Training and Simulation Market: All Set to Experience an Exponential Growth', Institute of Defence Studies and Analysis, 2009, http://www.frost.com/sublib/display-market-insight.do?id=178013878

102. Smith, Christopher R, 'Network Centric Warfare, Command, and the Nature of War', *CGSC*, 2009, http://cgsc.cdmhost.com/utils/getfile/collection/p4013coll3/id/2489/filename/2490.pdf

103. Smith, Edward R, 'Effect Based Operations', http://www.dtic.mil/dtic/tr/fulltext/u2/a457292.pdf

104. Smith, Philip, 'How Seriously Should the Threat of Cyber Warfare be Taken?' 2014, http://www.e-ir.info/2014/01/17/how-seriously-should-the-threat-of-cyber-warfare-be-taken/

105. 'Space Security: Fact Sheet', Space Security Organization, http://www.spacesecurity.org/SpaceSecurityFactSheet.pdf

106. Survey of India, http://www.surveyofindia.gov.in/

107. 'Technology Requirements of the Indian Army', http://www.ciidefence.com/pdf/Future_Technology_Requirements_of_the%20Indian_Army.pdf

108. Tertrais, Bruno, 'The Demise of Ares: The End of War as We Know It', *The Washington Quarterly*, 2012, http://dx.doi.org/10.1080/0163660X.2012.703521

109. 'The History of Cloud Computing', *China Daily*, 2013, http://www.chinadaily.com.cn/cndy/2013-01/21/content_16146386.htm

110. 'The Implementation of Network-centric Warfare', Office of Force Transformation, Department of Defense, http://www.au.af.mil/au/awc/awcgate/transformation/oft_implementation_ncw.pdf

111. The Jaipur Observatory, http://cadrans-solaires.pagesperso-orange.fr/monde/jaipur/jaipur_uk.html

112. 'The Military Simulation and Virtual Training Market 2011-2021', *Visiongain*, 2011, http://www.visiongain.com/Report/654/The-Military-Simulation-and-Virtual-Training-Market-2011-2021

113. 'The Space-Based Laser Integrated Flight Experiment-Global Missile Defense in the Boost Phase', *Team SBL-IFX*, http://www.wslfweb.org/docs/SBLWP.pdf

114. The United States Army 1995 Modernization Plan. Force 21, 1995, http://oai.dtic.mil/oai/oai?verb=getRecord&metadataPrefix=html&identifier=ADA286744

115. Tuner, Bunyamin, 'Information Operations in Strategic, Operational and Tactical Levels of War – A Balanced Systematic Approach', http://www.dtic.mil/dtic/tr/fulltext/u2/a418305.pdf

116. U.S. Army Cyber Command, http://www.arcyber.army.mil/

117. 'Understanding Datums and Projections', *Land Information*, http://www.linz.govt.nz/geodetic/find-out/understanding-datums

118. Villalpando, Carlos; Rennels, David and Some, Raphael, 'Reliable Multicore Processors for NASA Space Missions', http://trs-new.jpl.nasa.gov/dspace/bitstream/2014/41793/3/11-0119F.pdf

119. Warne, Leoni; Bopping, Derek and Ali, Irena, 'NetworkER Centric Warfare: Outcomes of the Human Dimension of Future Warfighting Task', 2006, http://www.dsto.defence.gov.au/attachments/Warne,%20Bopping%20&%20Ali_NetworkER%20Centric%20Warfare%20-%20Outcomes%20of%20the%20Human%20Dimension%20of%20Future%20Warfighting%20Task.pdf

120. Watts, Barry D, 'The Maturing Revolution in Military Affairs (Pt 2)', Center for Strategic and Budgetary Assessments (CSBA), http://www.isn.ethz.ch/Digital-Library/Articles/Detail/?lng=en&id=162686

121. 'What are geographic coordinate systems', *ArcGIS Resources*, 1998, http://resources.arcgis.com/en/help/main/10.1/index.html#//003r00000006000000

122. Wilson, JR, 'Network Centric Warfare 21st Century', http://www.afcea.org.ar/publicaciones/wilson.htm

123. Winter, Donald C, 'Adapting to the Threat Dynamics of the 21st Century', *The Heritage Foundation*, 2011, http://www.heritage.org/research/reports/2011/09/adapting-to-the-threat-dynamics-of-the-21st-century

124. World Geodetic System 1984, http://www.oosa.unvienna.org/pdf/icg/2012/template/WGS_84.pdf

Index